A Handbook of
Environment Impact Assesment

Related Books

Titles	*Authors*
— A Text Book of Environmental Science	*Vidya Thakur*
— Agricultural Input and Environment	*S.P. Palaniappan*
— Alkali Wastelands Environment and Reclamation	*I.C. Gupta*
— Applied Microbiology (Agriculture, Environmental, Food and Industrial Microbiology)	*S.M. Reddy*
— Biodiversity and Environmental Biotechnology	*P. Dwivedi*
— Climate Change and Environment	*J. Sundaresan*
— Climate, Agriculture and Man Shrinking Biodiversity and Sustainability	*V.K. Sehgal*
— Disaster Perception and Management	*T.S. Chouhan*
— Ecology of Desert Environments	*I. Prakash*
— Environment at Crossroads Challenges and Green Solutions	*Arun Arya*
— Environment Impact Assessment for Wetland Protection	*V.S. Kulkarni*
— Environmental Design and Development	*H.S. Sharma*
— Environmental Engineering and Safety	*Sangeeta Raut*
— Environmental Management in Mining Areas	*N.C. Saxena*
— Ecology Workbook	*R. Mishra*
— Environmental Pollution and Environmental Management	*P. Dwivedi*
— Environmental Protection	*A.K. Thukral*
— Environmental Resources and Tribes of North Western India	*V. Sankaran*
— Essentials of Environmental Studies	*S.S. Sagwal*
— Geo-informatics For Combating Land Degradation and Desertification	*T.S. Chouhan*
— Human Impact on Desert Environment	*P. Narain*
— Hydrogeology and Environmental Geology of Sonar River Basin	*S.K. Jain*
— Industrial Pollution and Environmental Management	*R.K. Trivedy*
— Industrial Waste Waters and Environmental Pollution	*D.C. Joshi*
— Mining Environment Management Manual	*N.C. Saxena*
— Objective Environmental Studies and Disaster Mgmt.	*M. Prasanthrajan*
— Pollution Risk and Impact on Social, Economic and Environmental Sectors and Remediation	*Golam Kibria*
— Prayavaran Javiki: Environmental Biology	*M.M. Trigunayat*
— Text Book of Disaster Management A.K. Shrivastava	
— Questions and Answers in Environmental Science Practical	*R.K. Kamble*
— Wasteland Management and Environment	*A.K. Roy*
— Wildlife Ecology And Conservation (21st Century Biology And Agriculture: Textbook Series)	*M. Balakrishnan*

A Handbook of Environment Impact Assesment

V.S. Kulkarni | S.N. Kaul
National Environmental Engineering
Research Institute, Nehru Marg,
NAGPUR - 440 020

R.K Trivedy
Reader & In-charge
Pollution Studies Deptt.
Y.C. College of Science, KARAD- 415 110

SCIENTIFIC
PUBLISHERS

Published by:

SCIENTIFIC PUBLISHERS (INDIA)
5A, New Pali Road, P.O. Box 91
Jodhpur 342 001, India
E-mail: info@scientificpub.com
Website: www.scientificpub.com

Print: 2025

ISBN: 978-81-72332-99-0
eISBN: 978-93-88449-12-0

Printed in India

PREFACE

Much of todays' Environmental crisis is due to scientific materialism and rampant economic development without adequate care of the environment and knowledge of physical limits by ecological systems on economic activity. Many of our environmental problems are due to the reliance on overexploitation of Natural resources, faulty planning and execution of infrastructure and industrial activity' and inadequate emphasis on parameters of sustainable development.

All this is chaning now with the emergence of the concept of sustainable development and shift towards ecological determination so as'to achieve growth within hhe limits of ecosystems carrying capacity.

Environment Impact Ass6ssment is potentially one of the most valuable interdisciplinary and objective decision making tools with respect to alternate routes for development, process technologies and project sites. It is an ideal anticipatory mechanism allowing measures that ensure environmental compatibility in our quest for socioeconomic development.

Although Environment Impact Assessment of major developmental projects was undertaken in 1950s in several countries; it was US National Environmental Policy Act of 1970 which gave fillip to EIA activities all over the world. In India serious environmental review of developmental projects started in the early 80s and with the release of EIA Notification of Govt. of India on 27th Jan. 1994, Environmental Impact Assessment of certain categories of projects became mandatory.

This book addresses itself to thd development of a tool to provide scientific inputs to the process of EI31 of industrial projects as a decision making tool to the development and computerization of methods for impact identification, prediction, evaluation and mitigation.

The concept of EIA and available methodologies have been discussed at length. The book also presents extensive literature survey on EIA as well Bs various practices followed all over the world. The literature survey also analyzes documented literature on site selection criteria, screening of projects for EIA, EIA methodologies, impact quantification techniques and scope for enlarged role of project EIA.

The book also presents, incorporating the best features of various EL4 methodologies and prediction models, a user friendly PC based software package acronymed EIA-AID. The software is developed in quick basic 4.5. The software developed is validated through a real life case study.

It is hoped that the book shall be of immense help to the students and teachers of Environmental Science and Engineering and also to Environmental Consultants, Industries and several Government Departments.

January 01, 2002 **V.S. Kulkarni**

CONTENTS

CHAPTER – 1

INTRODUCTION

1.1 CONCEPT OF EIA

The sociocultural roots of the present environmental crisis lie in the paradigms of scientific materialism and economic determinism which fail to recognize the physical limits imposed by ecological systems on economic activity. The economies must expand within ecosystems which have limited regenerative capacities. Contrary to the neoclassical theory of continuous material growth, economic activities directly undermine the potential for development through over-exploitation of natural resources, and indirectly compromise future production through the discharge of residuals. The entrenchment with quantitative growth as a major instrument of social policy in thus quite paradoxical (Khanna, 1989).

The emergence of the concept of sustainable development in recent years has brought in the general realization that societal perceptions must shift towards ecological determinism so as to achieve qualitative growth within the limits of ecosystem carrying capacity. The carrying capacity based planning process, innovative technologies for enhanced material and energy effectivity of production and consumption, structural conomic change towards . less resource-intensive sectors, and preventive environmental management through increasingly interventionist policies are some of the strategies for reconciling developmental goals with ecological capabilities (Khanna and Kulkarni, 1992).

Environmental impact assessment is potentially one of the most valuable, inter-disciplinary, and objective decision-making tools with respect to alternate routes for development, process technologies and project sites. It is an ideal anticipatory mechanism allowing measures that ensure environmental compatibility in our quest for socio-economic development.

There is no universally accepted definition of EIA. Some commonly used definitions are:

- EIA consists in establishing quantitative values for selected parameters which indicate the quality of the environment before, during, and after the action (Heer & Hagerty, 1977).
- EIA is an assessment of all relevant environmental and resulting social effects which would result from a project (Battelle, 1978).
- EIA is an activity designed to identify and predict the impact on the biogeophysical environment and on man's health and well-being of legislative proposals, policies, programmes, projects and operational procedures, and to interpret and communicate information about impacts (Munn,1979).
- EIA is a process to identify, predict and to describe in appropriate terms the pros and cons (benefits and penalties) of a proposed development. To be useful, the assessment needs to be communicated in terms understandable by the community and decisign-makers and the pros and cons should be identified on the basis of criteria relevant to the countries affected (UNEP, 1980).
- EIA is the systematic examination of environmental consequences of projects, policies, plaqs and programmes. Its main aim is to provide decision-makers with an account of implications of alternative courses of action before a decision is made (Clark, 1984).
- EIA is a . formal study process used to predict the environmental consequences of the proposed developmental project, find ways to reduce unacceptable impacts, and shape the project so that it suits the local environment, and present these predictions and options to decision-makers (UNEP, 1988).

The definition proposed by Hear and Hagarty forms the basis of present research. EIA is a process that has the objective of providing decision makers with an indication of the likely consequences of their actions. This is achieved through impact identification, prediction, evaluation and mitigation stages which endeavour to answer the questions : what will happen as a result of the project? what will be the extent of the changes? do the changes matter? and what can be done about them? (UNEP, 1988).

Over the years, it has become increasingly clear that the environmental clearance of the project proposals by the competent authority is not the sole decision point in EIA. Decision making starts much earlier, may be at the stage of selection of raw materials and production process, and extends to feedback mechanisms in the form of environmental monitoring during the operation phase. In the past, attention has tended to focus on environmental clearance, and the importance of a well- integrated appraisal in the refinement of project proposals has largely been under-valued (Wathern, 1988).

However, EIA is no longer seen as an add-on process. The greatest contribution of EIA to environmental management is in reducing adverse impacts by devising measures, beyond those warranted by emission standards and environmental management regulations, through the application of tools such as impact identification and use of mathematical models for impact prediction before the project proposal is submitted for environmental appraisal. It is now generally accepted throughout the world that the benefits of project level EIA considerably outweigh its costs (Wathern, 1988; Biswas and Agarwala, 1992; Wood, 1994). In recent years, the outcome of EIA studies ranges from simple impact statements to changes in project sites, manufacturing processes, raw materials, engint~ttring designs, introduction of additional pollution control measures, landscaping, manpower training programmes, compensation for restoration of damaged resources, and off-site programs to enhance quality of life of the community likely to be affected due to proposed de-deropmental activity.

1.2 HIERARCHY IN EIA

The EIA studies for industrial projects could be categorized as site selection studies, rapid/comprehensive/regional EIAs, and carrying capacity studies.

Site selection studies involve evaluation of the alternative sites with respect to environmental and project attributes such as proximity to raw materials, infrastructural facilities, markets etc. The study aims at ranking of site alternatives for objective decision making.

Rapid EIA differs from **Comprehensive EIA** with respect to tiineframe for baseline data collection. Rapid EIA refers to assessment based on one season (3 months) monitoring whereas Comprehensive EIA relates to assessment based on three seasons (9 months) monitoring fbr baseline data. Rapid EIA facilitates decision-making in those situations where a fair amount of knowledge exists about the proposed site or the impacts of the proposed development. Rapid EIA also helps in identifying significant issues for comprehensive EIA.

Regional EIA relates to development in/of a regon based on three season data collection and addresses itself to the analysis of assimilative capacity of air, water and land components of environment. The scope of carrying capacity study is extended to the analysis of supportive capacity in the region with respect to resource availability/utilization, supplyldemand, infrastructure/ congestion, and assimilative capacity/residuals.

1.3 MAJOR ISSUES IN EIA

EIA has been regarded as both science and art, reflecting the concern related to technical aspects of appraisal and the effects of EIA upon the decision making process (Kennedy 1984). Further, decisions are often shaped rather than made or taken. Everyone invo- lved in policy formulation, planning and impact assessment studies as well as in public hearings, reviews, legislative/media debates is in fact playing a part in shaping decisions (Munn, 1979).

The design of an effective EIA methodoloq can be envisaged as the search for mechanisms to deal with issues generated by the need to juxtapose the planning and environmental clearance of project proposals. Some issues deal with technical matters such as impact identification and prediction while others relate to the management of information. The major issues in EIA process are :

- Identification of projects requiring EIA
- Identification of impacts to be assesseci
- Assessment of impacts
- Completion of an appraisal
- Enabling the objectivity in appraisal

The structure of an EIA process is dictated primarily by the need to accommodate these issues as illustrated in Figure 1.1 (Wathern, 1988).

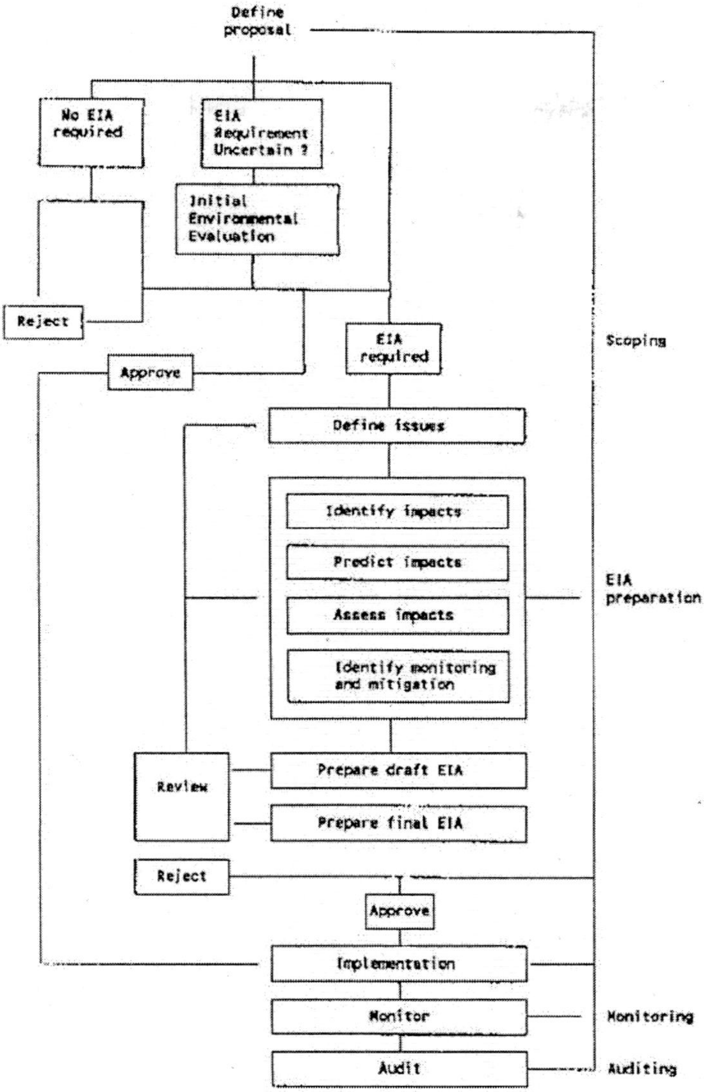

Fig. 1.1. Flow diagram showing the main components of an EIA system

From the technical point of view, EIA can be considered as a data management process. It has three components. First, the appropriate information necessary for a particular decision to be taken must be identified and, possibly, collected. Secondly, changes in environmental parameters resulting from the implementation of the project must be determined and compared with the situation likely to accrue without the project. Finally, actual change must be recorded and analysed.

The most important problem in EIA is the availability of appropriate EIA methodology. More than fifty impact assessment methodologies have been developed so far but no universal methodology is yet available for internalising environmental concerns in planning process.

Available methologies for EIA; which include checklists, matrices and networks; perform the task of identification of potential impacts. Mathematical modelling helps in prediction of impacts. Batelle Environmental Evaluation System and UNEP Test Model achieve the evaluation step. None of the methodologies, available singly, satisfactorily achieve the necessary three steps. Further, if essential to evolve a methodology that would lend itself to the integration of environmental concerns in the process of socio-economic development.

1.4 SCOPE FOR USE OF COMPUTERS IN EIA

Experience over the years has shown that EIAs are always conducted under severe limitations of time, manpower, financig resources, and data (Khanna, 1989).

In the past, the major role of computers in ELA was limited to the prediction step. However, computers can be used in all the stages of EIA, viz. impact identification, prediction, evaluation and mitigation.

The most logical way of identifying impacts associated with the project is through establishing cause-condition-effect networks. The complex nature of industrial processes warrant thorough knowledge of the transformation processes as also environmental risks associated with various activities in pinpointing the impacts of developmental projects and optimizing the data requirements of EIA.

Mathematical models for prediction of air, water and- noise environment are available in literature. However, the software packages incorporating these models have failed to keep pace with the modifications on the mathematical sciences. Further, these packages were not developed for EIA as such, are not user-friendly, and the out-come of the models need interpretation by experts. Most of the packages were developed in 1970s using FORTRAN, and the subsequent modifications have also relied on FORTRAN.

Scaling-weighting methods for impact evaluation facilitate aggregation of impacts to yield a score representing the net environmental impact of a project alternative. However, these methods warrant selection of parameters relevant to the project under consideration from a comprehensive list of parameters, computation of environmental quality for all selected parameters using value function curves, and assigning weights (parameter importance units) to the selected parameters through ranked pairwise comparison. Computational and data requirements of these methods have largely kept them at bay by the practioners of EIAs.

While the methodology for design of green-belt for attenuation of fugitive air emissions is very well documented, it is not commonly used due to complex mathematics involved in optimizing the width of green-belts. Since early 1980s, with the advent of personal computers (PC-XT to Pentium, monochrome to SVGA monitors) on one hand and availability of computer languages such as BASIC, Pascal and C which enable software development in interpretation mode and use of colours for screen display on other, computers have opened a wide vista to overcome the problems faced by the practioners of EIAs.

1.5 SCOPE OF CURRENT RESEARCH

The proposed research addresses itself to the development of tool a to provide scientific inputs to the process of EIA of industrial projects as a decision making tool through the development and computerization of methods for impact identification, prediction, evaluation and mitigation. The work has been organized into four parts:

- Literature review to screen available methodologies and software packages
- Development of methodology for EIA, and computer software package EIA-AID on Personal Computer platform

- Development of sensitivity analysis to assess uncertainty in impact prediction and evaluation, and con~puterizationo f the same
- Validation of EIA-AID through a case study

The literature review traces the evolution of EIA worldwide and in India, analyses documented literature on site selection criteria, screening of projects for EIA, EIA methodolo,aies, impact quantification techniques, and the scope for enlarged role of project level EIA.

Incorporating the best features of various EIA methodologies and prediction models a user-friendly PC based software package, acronymed EIA-AID, has been developed in Quick Basic 4.5 in this research to assist EIA teams. Salient features of EIA-AID are :

- The screening module enables selection of projects for rapid/comprehensive EIA based on the project threshold database developed during this research and inbuilt in the programme. The module also enables preliminary screening of site options based on siting guidelines developed in this research.
- The site selection module enables ranking of project site alternatives based on the considerations of regional assimilative capacity, supportive capacity, and project attributes. The methodology developed in this research, and incorporated in the module, uses the concept of scaling-weighing of impacts for ranking of site alternatives. The user has the options to either use in-built scaling functions and weights or define project specific ones in an interactive mode. Weights are assigned through an interactive ranked pairwise comparision technique. The structural mode of programme also facilitates scaling function database modification using the text-editor.
- The impact identification module incorporates the best features of questionnaire and descriptive checklists as also the concept of cause-condition-effect networks. The module enables identification of primary, secondary, and tertiary impacts trigerred by the project activities. The data requirements, sources of data, and tools for data analysis are also included. Adequate guidance is provided to the user on stanclnrcl practices of environmental monitoring.
- Thc prediction odules for air, noise and water qu:tlity ehable genc:rntioti of scenarios of c.nvironmenta1 qudiit) after the project implementation.

- The air quality prediction module provides options to select screening model for design of monitoring networks, shortterm predictions (1 hr, 8 hr and 24 hr average), sequential modelling, and long-term impact predictions based on stability rose data. The module incorporates state-of-the art versions of Gaussian Plume Model, wind extrapolation formulae, plume rise formulae, dispersion coefficients (urban/ rural areas), and terrain variations (flat/complex/valley). These models handle upto 50 sources at a time provide online guidance for input data, and generate output that does not warrant expert- interpretation. Salient features of the module lies in enabling the user to select different spatical resolutions in the impact zone (distance between a grid of 480 receptors located in 16 wind directions).
- The noise prediction module enables assessment of impact of stationary noise sources on a grid of 900 receptors defined in cartesian co-ordinate system. In adition, the user can specify the location of sensitive receptors (maximum 10 nos.). The module takes into account attenuation due to air absorbtion, physical barriers (defined by the user in two dimensions), and vegetation. The module outputs the impact of individual as also combined effect of noise sources. Both equivalent and day-night noise levels are computed. The module also computes the width of green-belt to attenuate noise levels.
- The water quality prediction module incorporates the self-calibrating model developed in the present research, andenables assessment of impact of carbonaceous and nitro - genous BOD on dissolved oxygen, impact of non- conservative pollutants with first order dacay coefficients, and conservative pollutants for riverststreams. The most significant features of the module lie in availability of deoxygenation and reaeration coefficients reported in literature, computation the coefficients based on selection of appropriate in-built formulae, and calibrates coefficients based on field data. In addition, it provides choice to over-ride all these computations to select coefficients based on user's judgement for with project scenario where these coefficients may change significantly based on site-specific conditions and waste characteristics.
- The impact evaluation nlodule incorporates a methodology developed in the prsent research by extending the concepts of Leopold matrix ancl Battcle Environmeutiil Evaluation System (BEES). The impact of' project is quantified in terms

of changes in ecological, pollution, aesthetics, and socioeconomic indices. These indices are then converted in numerical scores for assessing the overall impact.

- The sensitivity analysis module enables the user to assess the subjectivity in assigning weights in impact evaluation to delineate maximum/minimum weights that could be assigned to an individual environmental category/component/ parameter.
- The green-belt module enables optimization of the width of green-belt for abatement of fugitive emissions with different tree'species and atmospheric stabilities.

In order to assess the uncertainty in impact prediction, sensitivity analyses have been carried out for :

- Design of air quality monitoring network for multiple point sources
- Changes in maximum ground level concentrations of air pollutants due to :
  - changes in mixing height variations
  - changes in dispersion coefficients as a function of terrain variations
  - changes in wind extrapolation factors as a function of terrain variations
  - sequential analysis (averaging period 24 hours in calendar day vs. any 24 hours slot in study period)
- Impact zone for industrial noise sources
- Changes in dissolved oxygen deficit in receiving body of water due to :
  - variations in longitudinal dispersion
  - variations in deoxygenation coefficient
  - variations in reaeration coefficient

The software EIA-AID, developed in the present researcb, has been validated through a real-life case-study.

REFERENCES

Biswas, A.K., and Agarwala, S.B.C. (1992). Environmental Impact Assessment for Developing Countries, Butterworth-Heinemann, Oxford

Battele Institute, (1978). The Selection of Projects for EIA, Commission of the Europian Communities Environment and Consumer Protection Service, Brussels

Clark, B. D., (1984). Environmental Impact Assessment : Scope and Objectives in Perspectives in Environmental Impact Assessment, D. Reidel Publishing Company, Holland

Heer, J. E. and Hagerty, D. J. (1977). Environmental Assessment and Statements, Van Nostrand Reinhold, New York

Kennedy, W. V. (1984). US and Canadian Experience with. Environmental Impact Assessment: Relevance of the Europian Communities?, Zeitschriftfur Umweltpolitik, 7, pp. 339-66.

Khanna, P. (1989). Sustanable Development, Journal of Indian Association for Envir~nmentaMl anagement, Vol. 16, No. 1, pp. 6-19

Khanna, P. and Kulkarni, V. S. (1992). Policy Options for Environmentally Sound Technology, Environmentally Sound Technologies for Sustainable Development, ATAS Bulletin, Issue No. **7,** Spring 1992, United Nations, New York, pp 127-133

Munn, R. E. (1979). Environmental Impact Assessment : Principles and Procedures, Scope 5 Report, 2 nd edition, John Wiely, Chichester, England

UNEP, (1980). Guidelines for Assessing Indutrial Environmental Impact and Environmental Criteria for Siting of Industry, UNEP Industry and Environmental Guidelines Series, Vol. 1, UNEP Industry and Environment Office, Paris

UNEP, (1987). Environmental Law Guidelines and Principles : Environmental Impact Assess~entU, nited Nations Environment Programme, Nairobi

UNEP, (1988). Environmental Impact Assessment : Basic Procedures for Developing Countries, United Nations Environment Programme, Nairobi

Wathern, P. (1988). Introduction in Wathern, P. (ed.) Environmental Impact Assessment : Theory and Practice, Unwin Hyman, London

Wood, C. (1994). The Environmental Assessment of Plans rogrammes and Policies : A Comparative Review, 'Indo-British Workshop on Environmental Impact and Risk Assessment of Petrochemical Industry and Environmental Audit, January 8-10, 1994, Nagpur, India

CHAPTER – 2

LITERATURE REVIEW

2.1 EVOLUTION OF EIA WORLDWIDE

United States of America was the first country to assign mandatory status to EIA vide its National Environmental Policy Act (NEPA) of 1969. **A** host of industrialized countries have since implemented EIA procedures. Canada, Australia, the Netherlands and Japan adopted EIA legislation in 1973, 1974, 1981 and 1984 respectively. In July 1985 the European Community (EC) issued a directive making environmental assessments mandatory for certain categories of projects (Wood, 1994).

Among the developing countries, Columbia was the first Latin American Country to institute a system of EIA in 1974. In Asia and Pacific region; Thailand and Philippines have long- established procedures for EIA. EIA was made mandatory in Sri Lanka in 1984. A manual for EIA was prepared and published in 1988 in Sri Lanka (NBRO, 1988). There is dearth of information on the general situation in Africa, although a number of nations including Rwanda, Botswana and Sudan have experience of EIA (Wathern, 1988).

Bilateral and Multilateral agencies have also recognised the value of EIA as a decision-making tool. The Organization for Economic Co-opc!ration and Development (OECD) issued recommendations on EIA in its constituents states in 1974 and 1979, and fbr dcvt>lopment aid projccts in 1986. OECD issued guidelines for good practiccbs in **EIA** in 1992 (OECD, 1992). United Nations Environment I'rogrammc (UNEP) in 1980 provided guidance on EIA of' t h c~lcv c~lopmc~pntr oposals (UNEP, 1980) and supported research on EIA in developing countries (Ahmad and Swamy, 1985). UNEP, in 1987 vide the decision 14125 of the Governing Council, set out goals and principles of EIA for the member countries (UNEP, 1987) and provided guidance on basic procedures for EIA in 1988 (UNEP, 1988).

The World Conservation Strategy pinpointed the need to integrate environmental considerations with development in 1980 (IUCN, 1980). EIA. has become an integral part of World Bank policy (1987), which states that environmental issues must be addressed as part of overall economic policy. In 1989, the World Bank issued the Operational Directive on Environmental Assessment (O.D. 4.00) which was revised and updated in October 1991 (O.D. 4.01). Asian Development Bank in 1990 published guidelines for EIA (ADB, 1990). Importance of EIA was echoed in the Brundtland Report (WCED, 1987), and at United Nations Earth Summit on environment and development held at Rio de Jancrio in 1992 (UNCED, 1992).

As foreseen by Garner and O'Riodan (1982) development of **EIA,** as a tool for decision making world-over, has emerged through fOllowing stages :

- No formal accounting; decisions made on interest group lobbying and engineering feasibility; primary emphasis on economic development
- Conventional cost-benefit analysis; emphasis on efficiency criterion and engineering feasibility; major concern still economic development
- Innovative cost-benefit analysis, use of multiple objectives and discount rates, imaginative proxy pricing mechanisms; economic development as one of the objectives
- EIA mainly concerned with describing the repercussions of the proposals on bio-physical processes; economic development still primary objective
- EIA with more attention given to socio-cultural as well as bio- physical systems, economic development principle but not the sole objective
- EIA at two tier level, first to look at generic issues associated with the large projects'as a class, second to investigate how an "approved" scheme can best be designed with minimal net social and bio-physical disruption. EIA visualized as a creative participatory environmental management activity.

The summary of evolution of EIA in various countries is presented in Table 2.1 (Htun, 1988, Wathern, 1988, Wood, 1994).

Table 2.1. Evolution of EIA in various countries

Country	Basis of Im&mentation
Australia	Environmental Protection (Impact of Proposals) Act, 1974, Commonwealth of Australia
California	California Environmental Quality Act (CEQA) of 1971
Canada	Federal Environmental ,Assessment and' Review Process Guidelines Order 1984, Canada
China	Environmental Protection Law, 1979
Common wealth	Environment Protection (Impact of Proposals)
France	Law on Protection de la Nature, 1978
India	Notification dated May 5, 1994 under the Environment Protection - Act, 1986
Japan	Principles for Implementing EIA by Environment Agency, 1984
New Zealand	Resource Management Act 1991, New Zealand
Sri Lanka	National Environmental Act 1980, amended in 1986
Thailand	Improvement and Conservation of National Environmental Quality Act 1975, amended in 1978.
The Netherlands	EIA Policy, 1986
United States	US National Environmental Policy Act of 1969
United Kingdom	Environmental Assessment Guidelines, 1989
Western Australia	Environmental Protection Act 1986, Western Australia
West Germany	Cabinet resolution, September, 1975

2.2 EVOLUTION OF EIA IN INDIA

Environmentai Impact Assessment (EIA) of developmental projects was first introduced in India in 1977-78 when Planning Commission advised the then Department of Science & Technology to take up environmental appraisal of river valley projects. Subsequently, projects in mining, industry, thermal power, port and harbour sectors were brought under the purview of EIA. At present, the Ministry of Environment and Forests (MEF) is the nodal agency for EIA in India.

The MEF published guidelines for preparation of .EIA statements along with questionnaires and checklists for industrial project (MEF, 1985). Project proponents are required to provide relevant information as indicated in the guidelines alongwith the Environmental Impact Assessment Statement/Environmental Management Plan (EMP). A preliminary scrutiny of the project proposals in made by the technical experts of MEF. After ensuring that the main aspects are covered, EIA reports are placed before

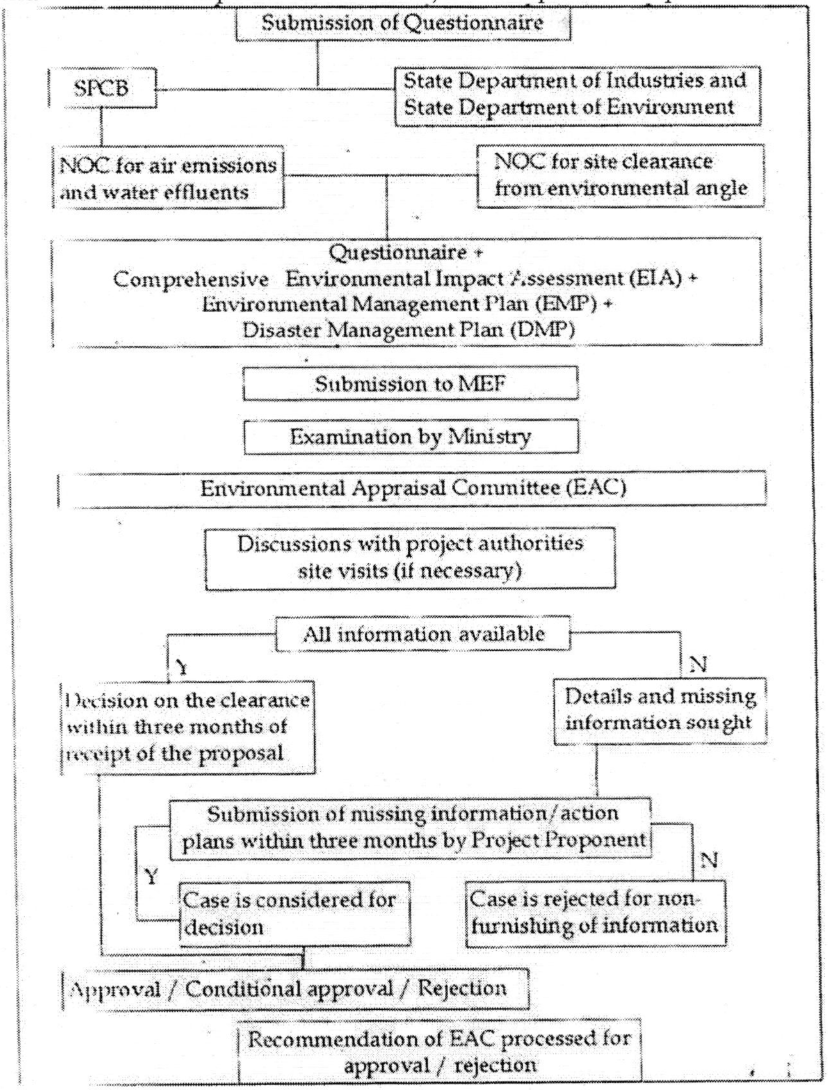

Fig. 2.1. Procedure for Environment Appraisal

the Appraisal Committees of Experts. The Appraisal Committees discusses the impact of the project with the project proponents and, if necessary, site visits are made for on-the-spot assessment of vari,ous environmental aspects. Based on their examination, the Appraisal Committees make recommendations on approval or rejection of the proposal by MEF. The environmental clearance to development projects is subject to implementation of stipulated safeguards under the provisions of Environment (Protection) Act 1986, Forest (Conservation) Act 1980, and other rules and regulations in force. Procedure for environmental appraisal of industrial projects in India is presented in Fig. 2.1.

The MEF on January 27, 1993, brought out draft notification under the Environment (Protection) Act 1986 making environmental clearance mandatory for expansion or modernization of any activity if pol'lution load is to exceed existing levels, and also for new projects listed in schedule - I of the notification. The notification was subsequently modified in May 1994 (MEF, 1994).

Over the years 1989-94,1286 EIAreports have been appraised at MEF (MEF, 1990, 1991, 1992, 1993, 1994). Statistical analysis of the same is pres'ented in Table 2.2. Analysis of EIA appraisal by MEF for industrial projects is presented in Fig. 2.2.

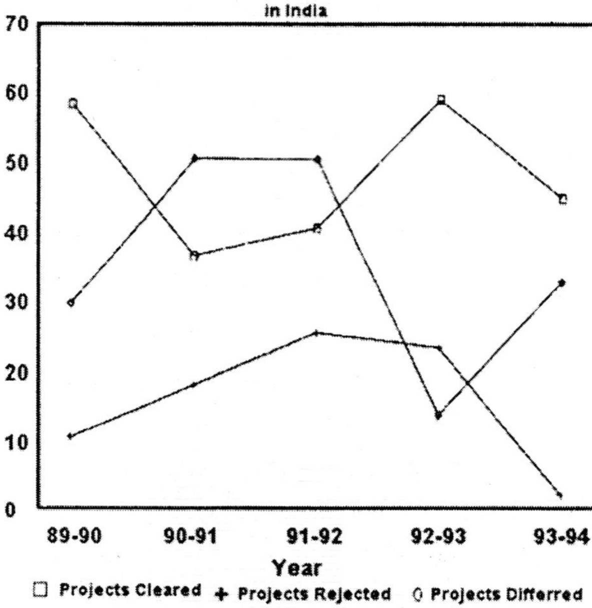

Fig. 2.2. EIA of industrial projects

Literature Review 17

Table 2.2. Environmental apperaisal of Development Project in India by IBP

Sector	Total no. of Projects Appraised						Projects Cleared						Projects Rejected						Additional Information Sought					
	89-90	90-91	91-92	92-93	93-94	To-tal	89-90	90-91	91-92	92-93	93-94	To-tal	89-90	90-91	91-92	92-93	93-94	To-tal	89-90	90-91	91-92	92-93	93-94	To-tal
Mining	69	75	51	55	70	340	7	20	21	20	42	109	48	20	12	2	4	82	34	34	35	14	33	149
Industry	53	59	70	52	37	271	31	22	29	31	27	130	6	10	14	13	2	45	16	27	27	27	6	96
Thermal power	61	40	37	20	24	182	28	17	16	13	5	83	9	11	13	0	3	34	24	12	12	4	7	65
Atomic power	0	10	6	3	1	10	0	4	3	2	2	10	0	1	0	0	6	1	0	5	5	3	1	9
River Valley	36	45	21	26	28	156	10	15	3	5	13	45	22	26	12	12	12	83	4	4	4	4	9	27
Other	120	61	39	51	54	325	67	30	24	10	24	171	20	23	2	2	2	59	33	2	2	27	19	95
To tal	359	290	224	247	314	1996	143	181	92	101	105	549	105	97	53	29	20	304	111	65	79	77	89	442

2.3 CLASSIFICATION OF ENVIRONMENTAL IMPACTS

The term 'impact' and 'effect' are frequently used synonymously in EIA literature, although some researchers have advocated differentiating between natural and man-induced changes, in the biogeophysical environment, effects, from the consequences of these changes, i.e. impacts (Munn, 1979). An impact has both spatial and temporal components and can be described as the change in. an environmental parameter over a specified period and within a defined area resulting from a particular activity compared with the situation which would have occurred had the activity not been undertaken.

Impacts can be classified as (Rau and Wooten, 1980) :

- adverse or beneficial
- caused or induced
- primary / secondary / tertiary
- short term / long term
- reversible / irreversible
- site specific / project specific

In view of the spatial and temporal distribution of the environmental impacts of industrial projects, Agrawal (1992) suggested that the impact of industrial development should be analysed in terms of host, proximate, significantly affected, moderately .affected, and slightly affected/unaffccted area. He further described the indicative geographical limits and nature of impacts (Table **2.3).**

2.4 PROJECT SCREENING

2.4.1 Methodologies for Screening of Projects

EIA procedure needs be appplied only to those actions which lead to significant environmental consequences. It is important to establish mechanisms for the selection of actions requiring EIA (Tomlinson, 1984). Screening allows elimination from the review process of all those projects that do not require detailed EIA, thus avoiding unnecessary expense and delays in project clearance.

Screening methods are based on :

- project thresholds
- positive and negative lists
- matrices
- initial environmental examinations

Table 2.3. Impact zone of industrial projects

Category	Geographical Limits		Nature and Magnitude of impact		
	Land Area	Stream strech	General	Air Quality Degradation	Water quality
A. Host	MajorIndustry : One Km/all around the battery limits. Medium Industry/indul. Estate: 0.5 km. all around periphery	1.0 km U/S of otfall to 5.0 km. D/S of oufall. in Case of lake/sea 1.0 km radius otfall.	Noise, traffic odours, dust disruption of socio-cultural life; grave imm-ediate and all time risk from accidents such as damage to life, health flora, fauna, property etc.	Servers at many times: e.g. cement plants SPM : 800-1500 μg/m3 Thermal Power/ Fertilizer Rfineries NH3: 2-5 PPM; SO2 : 150-300 μg/m3	Severs when diluion is low: ecology affected: fish kills may occur periodically
B. Proximity	Major industry: down wind 3 km. Others 2 km. Medium/indl. estate dominent down wind : 2kms o thers : 1 km	From 3 km. U/S upto 20 km D/S or where dillu-tion available becomes 20 times or more also 3 kms U/S on all tribuary streams joining in this reach.	Risk and damage when transient/accidental rel-eases accompany adverse meterological conditions, leaf injury in plants, health and life loss in humans/animals etc.	Server at some times: SPM : 500-800 μg/m3 SO2 : 100-150 μg/m3 NH3 : 1-2 PPM	As above Dissolved oxygen may touch 2.0 or lower at times, ecology may get severely affected
C. Significantly affected	Major industry : Domin-ant down wind : 5 kms Others: 3 kms. Medium/ Indl. Estates Dominant DW 3 kms Others : 2 Kms	Upto 40 kms. D/S or where diluion become 50 times or more	Nuisance on several occassions: effect on yield and growth of sensitive plants and health of sensitive humans and animals, risk in case of very serious accidents	Moderate occassionally SPM : 300-500 μg/m3 SO2 : 30-100 μg/m3 NH3 : 0.01-0.5 PPM	Dissolved oxygen recovers bu ecology still may be severly affected

D. Moderately affected	Major industry : Dominant down wind : 8 kms Others: 4 kms. Medium/ Indl. Estates DW 5 kms Others : 3 kms	Upto 80 kms D/S or where dilluion become 100 times or more	Some nuisance occassions, some risk, some subtle effects on flora and fauna of sensitive species and individuals	Some times SPM : 200-300 µg/m3 SO_2 : 20-30 µg/m3 NH_3 : 0.02-0.1 PPM	Ecology slowly recovering
E. Slightly affected unaffected	Beyond (d)	Beyond (d)	Imperceptible	SPM : <200 µg/m3 SO_2 : 10-20 µg/m3	Almost normal

Projects thresholds method relies upon the establishment of threshold values for key features of the project or its environment.
If a threshold is exceeded then EIA is required.

Reliance upon one threshold may give rise to an incorrect decisions. It is, therefore, a normal practice to link a series of different thresholds together, Thresholds, while being simple, may need frequent revision particularly in the light of experience, and inflation when financial criteria are used.

Positive lists are compiled through a review of existing developments, identifying those projects that eve rise to significant environmental damage. Lists are one of the simplest approaches to screening. The preparation of lists involves one major problem, viz. individual projects of the seme general type or class may have considerable variations in size of plant, process and layout which may give rise to varying environmental consequences. In addition, gaining the acceptance of all parties on the entry of individual project types on particular lists (mandatory positive, intermediate and negative) may be difficult and time consuming.

Project screening criteria, based on thresholds and lists, advocated by various countrieslagencies are summarised in Table 2.4 (OCCD, 1986, World Bank, 1991, Go, 1990).

The United Nations Environmental Programme (UNEP,1980) as also the Asian Development Bank (ADB, 1990) have developed questionnaires for the assessment of industrial activities, termed as Initial Environmental Examination (IEE), which provide a simplified decision tool for screening of projects for EIA. Guidelines for site screening are also published by UNEP (1980).

2.4.2 Project Screening Criteria in India

Notwithstanding the fact that EIA of industrial projects in carried out in India since 1984, no explicit screening criteria were available in India till 1994. The projects appraised by the MEF included (MEF, 1993) :

- Projects which require approval of the Public Investment Board/Planning CommissionlCentral Electricity authsriey
- Projects referred to MEF by other Ministries
- Projects which are sensitive and located in environmentally degraded areas
- Projects of Public sector undertakings of the Government of India where the project cost is more than Rs. 20 crores

The MEF, vide its notification of May 1994, has made environmental clearance mandatory for expansion or modernization of any activity if pollution load exceed existing levels, and also for new projects listed in Schedule - I of the notification (Table 2.5).

Table 2.4. Select Screening Criteria for Industrial Projects

I. Thailand

Petrochemical	:	Raw material requirement 100 TPD or more
Oil refining	:	All
Natural gas	:	All
Chlor-alkali	:	production capacity 100 TPD or more
Iron and steel	:	Production capacity 100 TPD or more
Cement	:	All
Smelting	:	Smelting capacity 50 TPD or more
Wood pulp	:	Production capacity 50 TPD or more

11. The Philippines

- Non-ferrous metal industrie.
- Iron and steel mills
- Petroleum and motor-chemical industries including oil and gas
- Smelting plants
- Power plants

111. Sweden

- Iron and metal works
- Large saw mills
- Factories producing organic and inorganic chemicals
- Plants processing nuclear fuels
- Oil refineries
- Fossil fuel power plants exceeding 500 MW
- Fertilizer
- Cement

N. Commission of Europian Communities

Projects to be subjected to mandatory EIA

Crude oil refineries (excluding undertakings manufacturing only lubricants from crude oil) and installations for gasification and 1iquificGion (greater than 500 tonnes or more of ccal or bituminous shale per day)

- Thermal power stations and other combustion installations (with a heat output of 300 MW or more)
- Integrated works for the initial melting of cast-iron and steel
- Installations for the extraction, processing and transformation of asbestos and products containing asbestos (installations for asbestos-cement products-annual production greater than 20,000 tonnes of finished product, for friction materials - greater than 50 tonnes of finished product per annum, and for other uses of asbestos-more than 200 tonnes per annum)

  Integrated chemical installations

Projects to be subjected to EIA when appropriate (Member State to decide on case to case basis)

- Glass
- Chemical
- Food processing
- Textile
- Leather
- Pulp and paper
- Rubber

V. OECD (1986)

- Chemical plants
- Power plants
- Industrial processes generating toxic wastes
- Pulp mills

VI. The World Bank

Category A : projects which affect the environment significantly through direct pollutant discharges that are large enough to cause degradation of air, water or soil; large scale physical distaurbances of the site and/or surroundings; extraction, consumption, or conversion of substantial amounts of forest and other natural resources; hazardous materials in more than incidental quantities; and involuntary displacement of people and other significant social disturbances. Large industrial plants and plants involving manufacture, transportation, and use of pesticides and other hazardous and/or toxic materials are included in Category A: projects. These projects require full environmental assessment. Category B : projects often differ from Category A projects of the same type only in scale. These projects require limited environmental assessment. Category C : projects do not require environmental assessment.

The final decision on classification of project rests with the Task Manager.

Table 2.5. Project Requiring Environmental Clearance as per Schedule-I, **MEF** Notification

1	Nuclear power and related projects such as Heavy Water Plants, nuclear fuel complex, rare earths
2	River valley projects including hydel power, major irrigation and their combination including food control
3	Ports, harbours, airports (except minor ports and harbours)
4	Petroleum refineries including crude and product pipelines
5	Chemical Fertilizers (Nitrogenous and Phosphatic) other than single superphosphate)
6	Pesticides (Technical)
7	Petrochemical complexes (both Olefinic and Aromatic) and Petrochemical intermediates such as DMT, Caprolacturn LAB etc. and production of basic plastics such as LLPDE, HPDE, PP PVC
8	Bulk drugs and pharmaceuticals
9	Exploration for oil and gas and their production, transportation and storage
10	Synthetic rubber
11	Asbestos and asbestos produuts
12	Hydrocyanic acid and its derivatives
13	Primary metallurgical industries (such as production of iron and steel, aluminium, copper, zinc, lead and ferro alloys) Electric arc furnaces (mini steel plants)
14	Chlor alkali industry
15	Integrated paint complex including manufacture of reslns and basic raw materials required in the manufacture of paints
16	Viscose staple fibre and filament yarn
17	Storage batteries integrated with manufacture of oxides of lead and lead antimony alloy
18	All tourism projects between 200m-500m of high water line and at locations with an elevation of more than 1000 m with investment of more than Rs. 5 crores
19	Thermal power plants
20	Mining projects (with leases more than 5 hectares)
21	Highway projects
22	Tarred roads in Himalayas and or Forest areas
23	Distilleries
24	Raw skins and hides
25	Pulp, paper and newsprint
26	Dyes
27	Cement
28	Foundries (individuai!
29	Electroplating

2.5 METHODOLOGY FOR SITE SELECTION

The magnitude of environmental impacts of industrial projects is, in' most of the cases, site specific. Hence, the selection of site sllould form an integral part of EIA of industrial projects. However, there are no Aplicit methodologies documented in literature for sit.(? selection. Certain guidelines are provided in the Initial Environmental Examination format of ADB (1990) as presented in Table 2.6. The World Bank has advised its Task Managers to specifically consider project locations in the categorisation of the projects for further environmental assessment (World Bank, 1991).

In India, the Ministry of Environment and Forests (MEF, 1985) stipulated that the industrial sites should maintain the followingd istances from the areas listed below:

- Ecologically and ndlor otherwise sensitive areas : at least 25 kms, the requisite distance may be increased depending on the geo-climatic conditions
- Coastal areas: at least 1/2 km from high tide line
- Flood plains of the riverine systems : at least 112 km from flood plain or modified flood plain affected by dam in the upstream or by flood control systems
- Transport/communication systems: at least 112 km from highways and railways
- Major settlements (more than 3,00,000population) : distance from settlements is difficult.to maintain because of urban sprawl. At the time of siting of the industry if any notified limit of any major settlement is within 50 km, the spatial direction of growth of the settlement for at least a decade must be assessed and the industry shall be sited at least 25 km from the projected growth boundary of the settlement.

The list of ecologically or otherwise sensitive areas drawn by I hc. MEF is presented in Table 2.7.

The MEF has further stipulated that the following factors must be recognized in siting:

- No forest land shall be converted into non-forest activity for the industry (Forest Conservation Act, 1980)
- No prime'agricultural land shall be converted into industrial site
- Within the acquired land the industry must locate itself at the lowest location to remain obscured from general sight

Table 2.6. Checklist of Environmental Parameters for Industrial Projects Siting

Actions Affecting Environmental Resources and Values (A)	Damages to Environment (B)	R ecommended Feasible Pro tection Measures (C)	IEE (D)			
			No significnat Effect D1	Significant Effect		
				Small D2	Mode-rate D3	Major D4
A. Environmental Effects due to Project Location	A. Environmental Losses Unnecessarily High Because of Site Location	A. Consider Alternative Sites				
1. Proper site location	1.	1.				
2. Adeq uacy of buffer strip	2. Damages to neighborhood which could be avoided by proper buffer	2. Furnish appropriate buffer				
3. Excessive traffic hazards for access road	3. Losses in community econ omic and socio-econ omic values	3. Careful planning/operations/ monitoring to minimize and offset adverse effects				
4. Nuisances/hazards to neighbors	4.	4.				
5. Effects on adjacent property values	5.	5.				
6. Plant drainage problems	6. Flooding of plant and of neighborhood areas	6.				

7. Resettlement problems	7. Socia injustices	7.
8. Effects on precious ecology	8. Losses in precious ecology	8.
9. Socioecon omic impacts	9. Dissatistifed neighborhood population	9.
10. Water supply and effects on hydrology	10. Losses in water supply po tentials for ther uses	10.
11. Depreciation of environmental aesthetics by structure	11. Losses in environmenta aesthetics	11. Careful planning/operations/ monitoring to minimize and offset adverse effects, including screening and blending
12.Impairment of historical/ cultura mon uments/ values by structures	12. Loss in historical/cultura mon uments	12. Careful planning to minimize losses

- Land acquired should be sufficiently large to prov·de space for appropriate treatment of wastewater after maximum possible in-house recycle and reuse. Reclaimed (treated) wastewater shall be used for raising green belt and, if possible, for aquaculture
- The green belt between two adjoining large scale industries shall be one kilometer
- Enough space shall be provided for storage of solid wastes
- Layout and form of the indqstry must conform to the landscape of the area without affecting the scenic features at that location
- Associated township of the industry must be created at a space having phytographic barrier between the industry and the township.

Table -2.7. List of Ecologically and/or Otherwise Sensitive Areas

1	Religious and historic places
2	Archaeological monufients/sites
3	Scenic areas
4	Hill resorts
5	Beach resorts
6	Health resorts
7	Coastal areas rich in corals, mangroves, breeding grounds of specific species
8	Estuaries rich in mangroves, breeding grounds of specific species
9	Gulf areas
10	Biosphere reserves
11	National parks and sanctuaries
12	Natural lakes, swamps
13	Seismic zones
14	Tribal settlements
15	Areas of scientific and geological interests
16	Defense installations, specially those of security importance and sensitive **to** pollution
17	Bordera areas (international)
18	Aimorts

Ideally, the site for an industrial project should' be selected based on the regional carrying capacity (Khanna ***et*** *al.*, **1991)** which is defined as the maximum rate of resource consu1.1ption and waste discharge that can be sustained indefinitely in a defined planning region without progressively impairing bio-productivity and ecological integrity.

2.6 EIA METHODOLOGIES

2.6.1 Introduction

The identification and assessment of environmental impacts of developmental activities are complex because of the diversity of impacts caused by human interference to the environmental and social systems. Impact identification and assessment requires collection and collation of large amount of diverse data and, most importantly, communicatingth e final results to the decision makers and members of the public, many of whom are unlikely to be experts in environmental science. It is precisely to help overcome these difficulties that most EIA methods have been devised.

The need for systemization and interdisciplinary studies in EIA led to the development of common structured framework for relating different kind of data. Also, the task of comparing alternative project sites for industries demanded an assessment which allowes direct comparisons to be made between the alternatives. The variety of structured approaches to EIA developed to fulfill these tasks are generally known as EIA methodologies or methods (Bisset, 1984).

2.6.2 Review of EIA Methodologies

Although there is a large number of EIA methodologies which have been developed and used in EIA, the variety is more apparent than real. Nearly all methods are variants of general types which have specific organizing principles in common. These are checklists, matrices, networks, overlays and models. In case of checklists and matrices there are many versions whereas there are fewer examples of networks and models available. All these methodologies are generic in nature and are adapted to specific type of project assessment such as industrial, water resources, mininig, etc.

2.6.2.1 Checklists

The term checklist covers a variety of methods having widely varying characteristics and degree of complexity. The three basic categories of checklists are : simple checklists, questionnaire checklists and scaling-weighing checklists.

Simple Checklists

Simple checklists consist of a list of environmental, economic and social factors which may be affected by specific project actions. These are useful for impact identification. While such a checklist acts as a framework for identifying impacts, additional guidelines are needed to perform the other EIA tasks.

Descriptive Checklists

A comprehensive checklist accompanied by specific guidance on the data requirements, sources of information, and predictive techniques was developed by Schaenman (1976). The guidance allows this checklist to be utilized for impact identification, measurement, and presentation of results. It, however, cannot be used to develop and utilize means of determining the relative importance of different impacts.

Questionnaire checklists

Another t.ype of checklist is the questionnaire which prcscnts a series of questions relating to the ir.pacts of a project. E'ibwrc 2.3 presents the section on terrestrial ecosystems (there are 31 scctions in total) from a questionnaire checklist developed by the US Agency for International Development (1980). The method was designed for the assessment of rural development projects in dcvcloping countries. The questions are listed under generic categories such as terrestrial ecosystems and disease vectors. Those assessing projects must attempt to answer all the questions. Three answctrs are possible depending upon how much is known about a particular impact.

If insufficient information is available for a definite response, then unknown (Fig. 2.3) would be recorded. If an answer could be provided then it would be possible to use the scale for designating the expected impact. For example, an impact may be highly adverse or medium beneficial. The assessment of a project therefore consists of a systematic question and answer procedure with the addition of quantitative and qualitative information where appropriate.

A conceptually similar checklist is containcd in an assessment manual produced for bse by UK local authorities, central government, and developers (Clark *et al.*, 1981).

Scaling Checklists

The US Forest Service developed a checklist incorporating

a.	Are there any terrestrial ecosysterns of the types listed below which, by nature of their size, abundance or type, culd be classified as significant or uique?				
		Forest? Savanna? Grassland Desert?	Yes Yes Yes Yes	No No No NO	Unk. Unk. Unk. Unk.
b.	Are these ecosystems	Pristing? Moderately Degraded? Severely Degraded?	Yes Yes Yes	No No No	Unk. Unk. Unk.
c.	Are there present trends towards alteration of these ecosystems cutting, burning, etc. to produce agricultural, industrial or urban land?		Yes	No	Unk.
d.	Does the local population use these ecosystems to obtain non- domesticated:	Food plants? Medicinal plants? Wood products? Fiber? Fur? Food animals?	Yes Yes Yes Yes Yes Yes	No No No No No No	Unk. Unk. Unk. Unk. Unk. Unk.

Fig. **2.3.** Section from a Questionnaire checklists

Threshold of Concern (Sassaman, 1981). This checklist consists of a list of environmental elements or resources such as water quality, wildlife habitat, and air quality accompanied by criteria which express values called "Thresholds of Concern" of these resources which are desirable or which should not be reduced (Figure 2.4).

The elements, appropriate criteria, and Threshold of Concern values need to be selected for individual projects by those involved in the EIA. Threshold of Concern value for each environmental element is the point at which those assessing a proposal become concerned with the impact. Any impact that exceeds threshold of concern is considered to be significant to decision makers.

Impacts are considered in terms of their duration and reversibility, The checklist assumes that all impacts have equal importance.

Elemen t	Criteria	Threshold of concern (TOC)	Alternative 1 No action		Alternative 2 median investmen t		Alternative 3 high investmen t	
			Impact and duration	Impacts >10C?	Impact and duration	Impacts >TOC?	Impacts and duration	Impacts >10C?
Air quality	State guidelines	3	4C	Yes	4C	Yes	4C	Yes
Economics	Efficiency (benefit cost ratio)	1:1	3:1	No	4:1	No	4:5:1	No
Employmen t	Private sector jobs	Presen t level	9,000 C	No	9,500C	No	10,000C	No
	Forest service man-power requiremen ts	Presen t + 10%	400 C	No	440C	No	500C	Yes
Range resources	Animal unit mon ths provided	Presen t level	5,000 C	No	5,000C	No	3,000C	Yes
Recreation	Number of dispersed camping sites	5,000	2,800 C	Yes	5,000C	No	6,000C	No
	Win ter sports (visitor days)	1,000,000	700,000 C	Yes	1,000,000C	No	2,000,000C	No
Threatened and endangered wildlife species	Number of spotted owls (pairs)	35	50 D	No	35D	No	20 D	Yes
Water quality	State water quality standards	3	3C	No	3C	No	4C	Yes
Wildlife	Viewing deer and elk	25% decrease in population	10%C	No	10%C	No	30%C	Yes

Legend : A : Impact duration < 1 yr B : Impact duration 1-10 yrs
C : Impact duration 10-50 yrs D : Irreversible impact
3 : Classification of water body/area

Fig. 2.4. Threshold-of-concern Worksheet with Sample Elements, Criteria, Thrsholds of Concern and Impact data

Scaling-Weighing Checklists

Condensing information on cdmplex enviromental variables into some manageable form is a recurrent problem of assessment. Aggregation with scaling ahd weighing methods represent a technical fix to this problem (Dee *et al.,* 1973, Solomon *et al.,* 1977).

Scaling weighing checklists have been devised to enable all adverseheneficial impacts for a single project, or more usefully, alternative projects to be compared in the form of quantitative indices. The result has been the formulation of quasi-mathematical methods in which impacts are weighted in. terms of relative importance, transfqrmed into units from a common, notional scale and finally manipulated mathematically to form impact indices. The most well-known of these methods is the Battele Environmental Evaluation System (BEES) employed for water resources projects by Dee *et al.* (1973). Since then BEES has been adapted for EIA of water quality management plans, highways, nuclear power plants, and other projects.

BEES (Whitman, 1971) is based on scaling-weighting checklist consisting of a description of the environmental factors pertinent to the project alongwith the instructions for scaling the values of each parameter and assigning importance units. The structure of the BEES is hierarchical in nature to account for different levels of information used in the impact analysis. Four* levels of information are used in the BEES, viz.

Level 1 : Most general information- Environmental categories

Level 2 : Intermediate information - Environmental components

Level 3 : Specific information - Environmental parameters

Level 4 : Most specific information – Environmental measurements

BEES consists of a checklist of 78 parameters mostly covering the physical environment, although some social and economic factors are also included (Table 2.8) (Munn, 1979).

This method is essentially concerned with the determination of environmental quality of an area without a project, and then predicting a net positive or negative change, should the project be implemented. This involves predicting the future state of the environment without the project in the same time-scale as that used. to predict the project-affected environment.

Table 2.8. Battelle environmental evaluation system for Water Resource Development Development Projects

Ecology (240)	Terrestrial Species & Population (70) - Browsers and grazers (14) - Crops (14) - Natural vegetation (14) - Pest species (14) - Upland game birds (14) Aquatic Species and Populations (70) - Commercial fisheries (14) - Natural vegetation (14) - Pest species (14) - Sport fish (14) - Water fowl (14) Terrestrial Habitats & Communities (50) - Food web index (14) - Land use (12) - Rare & endangered species (10) - Species diversity (1 4) Aquatic Habitats & Communities (50) - Food web index (12) - Rare & endangered species (12) - River characteristics (12) - Species diversity (14)
Aesthetics (153)	Land (32) - Geologic surface material (6) - Relief & topographic character (16) - Width and alignment (10) Air (5) - Odour and visual (3) - Sounds (2) Water (52) - Appearance of water (10) - Land & water interface (16) - Odour and floating material (6) - Water surface area (1 0) - Wooded and geolog-lc shoreline (10) Biota (24) - Animals-domestic *(5)* - Animals - wild (5) - Diversity of vegetation types (9) - Variety within vegetation types (5) Man-Made Objects (10) - Man-made objects (10)

	Composition (30)
	- Composite effect (15)
	- Unique composition (15)
Physicd Chemical (402)	Water Quality (318)
	- Basin hydrologic loss (20)
	- Biochemical oxygen demand (25)
	- Dissolved oxygen (31)
	- Fecal coliforms (18)
	- Inorganic carbon (22)
	- Inorganic nitrogen (25)
	- Inorganic phosphate (28)
	- Pesticides (16)
	- pH (18)
	- Streamflow variation (28)
	- Temperature (28)
	- Total dissolved solids (25)
	- Toxic substances (14)
	- Turbidity (20)
	Air Quality (47)
	- Carbon monoxide *(5)*
	- Nitrogen oxides (10)
	- Particulate matter (12)
	- Photochemical oxidants (5)
	- Sulphur oxides (10)
	- Other (5)
	Land Pollution (28)
	- Land use (14)
	- Soil erosion (14)
	Noise Pollution (9)
	- Noise (9)
Human Interest/ Social(205)	EducationIScientific (48)
	- Archeological (13)
	- Ecological (13)
	- Geological (11)
	- Hydrological (11)
	Historical (55)
	- Architecture and styles (11)
	- Events (11)
	- Persons (11)
	- Religions and cultures (11)
	- Western frontier (11)
	Cultures (28)
	- Indians (14)
	- Other ethnic groups (7)
	- Religious groups (7)

MoodIAtmosphere (37)
- Awe/inspiration (11)
- Isolationfsolitude (11)
- Mystery (4)
- Oneness with nature (11)
Life Patterns (37)
- Employment opportunities (13)
- Housing (13)
- Social interactions (11)

Brackebd numbers are relative weights

The major feature of BEES is that environmental impacts are eqressed in commensurate units providing common units for comparison. The steps involved in the development of commensurate units include transformation of parameter estimates into an environmental quality (EQ) scale, assignment of importance units to the individual parameters (PIU), and the multiplication of scale values by importance values to obtain environmental impact units (EIU).

Environmental quality is measured by the use of value hctions for each of the 78 parameters thereby providing a numerical index of overall quality. To achieve this index it is assumed that the quality state of each parameter can be expressed on a scale of 0-1, where 1 represents highest quality, and 0 represents lowest quality. The value function for dissolved oxygen indicates that a level of 10 mgtl equals a quality score of 1, while 2 mgfl equals about 0.1. The determination of these quality scores for a variety of environmental components is termed scaling.

To assess the impacts of a project it is necessary to determine the relative importance of various parameters. This is achieved by using experts to assign numerical weights to each parameter. Thousand units are divided between the parameters according to their importance as determined by the experts. To obtain the numerical index for the overall quality, the weightage is multiplied by the quality score for each parameter. The summation of each would then indicate the overall quality state of that area. It is then necessary to predict the likely future state of each parameter assuming the project took place. This may be achieved by mathematical modelling or the use of expert opinion. When the total scores for with- and-without project situations are compared, the overall impact of the project can be measured. From such an analysis it would then be possible to select the least damaging alternative. Another benefit of this system is that it has a system of red flags by which serious impacts are highlighted.

Select value function curves used in BEES (Canter, 1982) are presented in Figures 2.5 through 2.7. Typical outcome of BEES application to EIA is presented in Table 2.9 (Canter, 1982).

Table 2.9. Trade-Off Analysis in the Batelle Environmental Evaluation System

Category	Maximum Score	Alternative A	A-ctual Score Alternative B	Alternative C
(240) Ecology	240	215	200	220
(402) Environmental pollutiop	402	340	350	310
(153) Aesthetics	153	110	120	135
(205) Human interest	205	180	175	175
Total	1000	845	845	840

Legend : Alternative A : no action .

The numbur in parenthesis is the maximum for a category

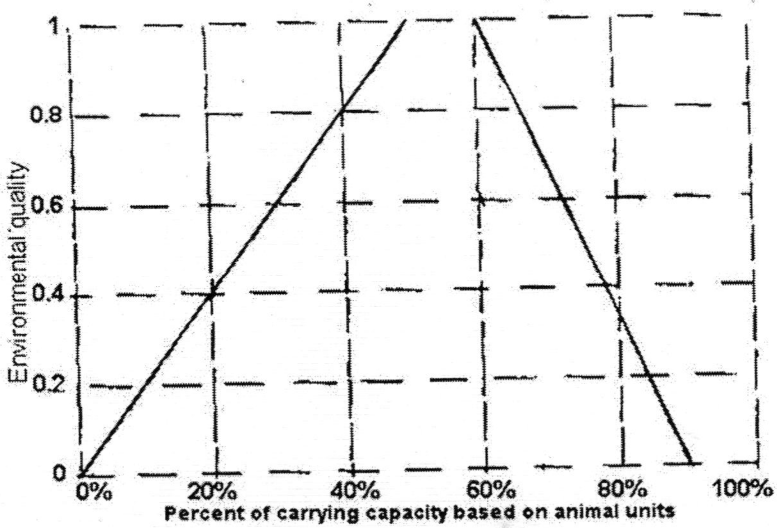

Fig. 2.5. Terrestrial carrying capacity for browsers and grazers

By developing value function curves for the socio-economic conditions BEES has- been adapted by workers in various countries

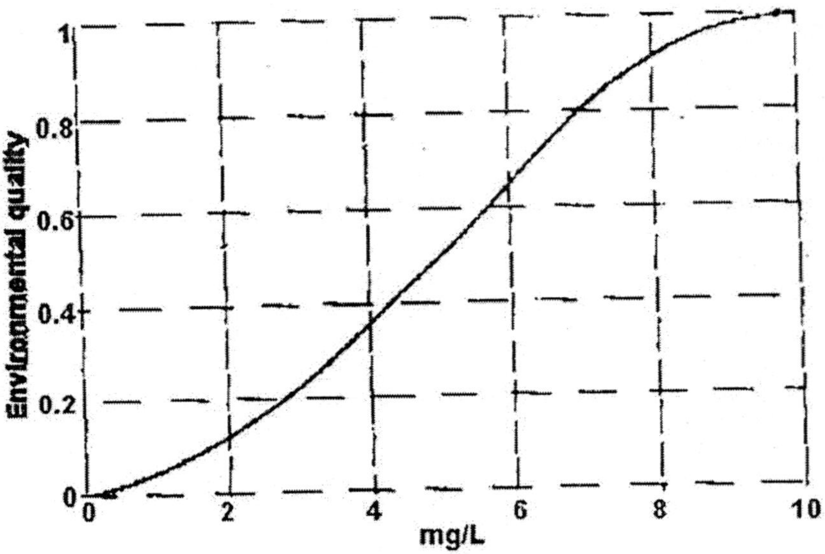

Fig. 2.6. Dissolved oxygen

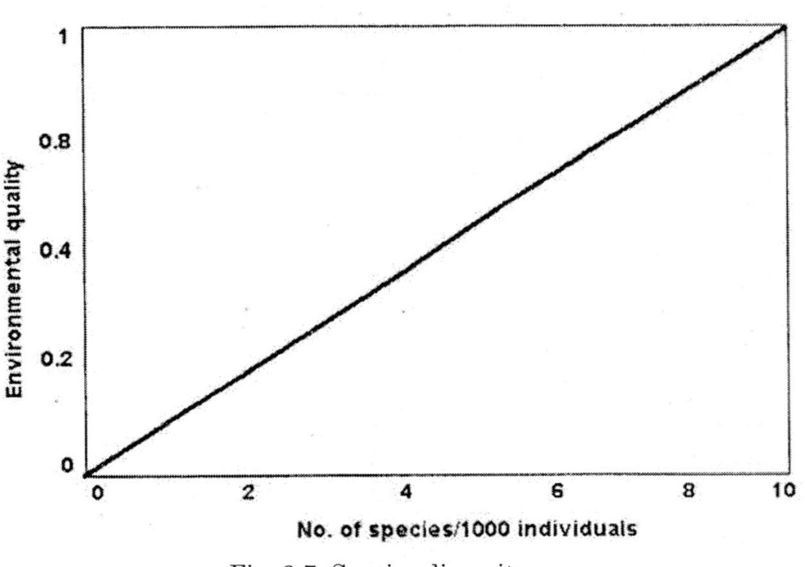

Fig. 2.7. Species diversity

(Bisset, 1984). Mangkol (1983) developed an adapted version of BEES through incorporation of impact magnitude (severity of impacts), prevalence (extent of an impact), duration and frequency (the time element), risk (probability of impact occurrence), importance (significance of impact at the time of EIA work), and mitigation (action that can be taken to reduce or eliminate adverse impacts). This variant of EES involves use of modified value functions to make them flexible to take account of bcal conditions. The modified value function incorporates haximum and minimum allowable levels for selected environmental parameters. Mangol uses the ratio of beneficial change in environmental impact units to adverse change in environmental impact units as the criterion for project appraisal (Bisset, 1984).

One of the major difficulties with scaling weighing checklists relates to the derivation of weights. Sondheim (1978) attempted to broaden the base of weight allocation, by removing it from those experts who are involved directly in specific EIA studies. In the method proposed by him weighingpanel is constituted for each EIA. Each member of the panel produces an individual weighing scheme which is used to produce a composite score for a number of alternative projects.

One important drawback of BEES is its focus on environmental parameters which can be quantified to the eytent that there is a great temptation to quantify the unquantifiable, eg. aesthetics. Further, subjective views incorporated in EES are representative of a very restricted population, namely experts (Bisset, 1984).

Many subjective elements are subsumed within both the weighing scheme and the value functions. In effect, the basis for decision is created by the method. The preferred option and, hence, the only decision that logically can be taken is one with the highest score for environmental quality.

BEES is complex to operate and any Environmental Impact Statement (EIS) prepared using this or a similar method would not be corhprehended easily by non-experts. Not only would the technical information on impacts needs to be understood, but also the principles of the method need to be comprehended. Finally, the complexity of this approach results in requirements of large manpower. time and resource requirements especially when determining the value functions.

Despite these drawbacks, BEES is regarded the most suitable

method to aggregate impacts of a developmental project and present them to the decision makers.

2.6.2.2 Matrices

Matrix methods are two-dimensional checklists in which possible project activities are established along one axis, with potentially impacted environmental characteristics or conditions along the other. Thus the cause-effect relationships between specific activities and impacts are identified in matrix methods.

Leopold et al. (1971) were the first to suggest the use of a matrix method for EIA. The Leopold Matrix is complex. A total of 8800 cells result from 88 environmental paramctcrs along one axis and 100 developmental characteristics along thc: other (Table 2.10). The matrix format is ideally suited for impact idc!ntificntion (Munn, 1979).

Table 2.10. Leopole Matrix

Part I : Project Actions

A. Modification of Regime	Exotic flora or fauna introduction
	Biological controls
	Modification of habitat
	Alteration of ground cover
	Alternation of ground-water hydrology
	Alternation of drainage
	River control and flow codification
	Canalization
	Irrigation
	Weather modification
	Burning
	Surface paving
	Noise and vibration
B. Land Transformation and Construction	Urbanization
	Industrial sites and buildings
	Airports
	Highways and bridges
	Roads
	Railways
	Cables and lifts
	Transmission lines, pipelines, and corridors
	Barriers, including fencing
	Channel dredging and straightening
	Channel revetments

	Canals
	Dams and impoundements
	Piers, seawalls, marines & sea terminals
	Offshore structures
	Recreational structures
	Blasting and drilling
	Cut and fill
	Tunnels and underground structures
C. Resource Extraction	Blasting and drilling
	Surface excavation
	Sub-surface excavation and rething
	Well drilling and fluid removal
	Dredging
	Clear cutting and other lumbering
	Commercial fishing and hunting
D. Processing	Farming
	Ranchign grazing
	Feed lots
	Dairying
	Energy generation
	Mineral processing
	Metallurgical industry
	Chemical industry
	Textile industry
	Automobile and aircraft
	Oil refining
	Food
	Lumbering
	Pulp and paper
	Product storage
E. Land Alteration	Erosion control and terracing
	Mine sealing and waste control
	Strip mining rehabilitation
	Lanscaping
	Harbour dredging
	Marsh fill and drainage
F. Resource Renewal	Reforestation
	Wildlife stocking and management
	Ground-water recharge
	Fertilization application
	Waste recycling
G. Changes in Traffic	Railway
	Automobiles
	Trucks
	Shipping
	Aircraft

	River and canal traffic
	Pleasure boating
	Trails
	Cables and lifts
	Communication
	Pipeline
H. Waste Emplacement and Treatment	Ocean dumping
	Landfill
	Emplacement of tailings, spoil, and overburden
	Underground storage
	Junk disposal
	Oil-well flooding
	Deep-well emplacement
	Cooling water discharge
	Municipal waste discharge including spray irrigation
	Liquid ef'fluent discharge
	Stabilization and oxidation ponds
	Septic tanks, commercial & domestic
	Stack and exhaust emission
	Spent lubricants
I. Chemical Treatment	Fertilization
	Chemical deicing of highways, etc
	Chemical stabilization of soil
	Weed control
	Insect control (pesticides)
J. Accidents	Explosions
	Spills and leaks
	Operational failure
Others	

Part II : Environmental Characteristics and Conditions (Impacts)

A Physical and Chemical Characteristics

1. Earth	Mineral resources
	Construction material
	Soils
	Landform
	Force fields & background radiation
	Unique physical features
2. Water	Surface
	Ocean
	Underground
	Quality
	Temperature, Recharge, Snow, **ice,** permafrost

3. Atmosphere	Quality (gases, particulates)
	Climate (micro, macro)
	Temperature
4. Processes	Floods
	Erosioll
	Deposition (sedimentation, precipitation)
	Solution
	Sorption (ion exchange, complexing)
	Compaction and settling
	Stability (slides, slumps)
	Stress-strain (earthquake)
	Air movements
B. Biological Conditions	
1. Flora	Flora
	Trees
	Shrubs
	Grass
	Crops
	Microflora
	Aquatic: plants
	Endangered species
	Barriers
	Corridors
2. Fauna	Birds
	Land animals including reptiles
	Fish & shelfish
	Henthic organisms
	Insects
	Xlicrofauna
	Endangered species
	Barriers
	Corridors
C. Cultural Factors	
1. Idand Use	Wilderness & open spaces
	Wetlands
	Forestry
	Grazing
	Agriculture
	Residential
	Commercial, Industrial, Mining & quarryin,
2, Recreation	Hunting
	Fishing
	Boating
	Swimlning

	Camping & hiking
	Picnicking
	Itesorts
3. Aesthetics & Human Inkrest	Scenic views and vistas
	Wilderness qualities
	Open space qualities
	Lanscape design
	Unique physical features
	Parks & reserves
	Monunlents
	Rare & unique species or ecosystems
	Historical or archeological sites and objects
	Presence of misfits
4. Cultural Status	Cultural patterns (life style)
	Health and safety
	Employment
	Population density
5. Man-Made Facilities and Activities	Structures
	Transportation network (movement, access)
	Utility networks
	Waste disposal
	Barriers
	Corridors
D. Ecological Itelationship Such as :	
	Salinization of water resources
	Eutrophication
	Iliseasc-insect vectors
	Food chains
	Salinization of surficial material
	Brush encroachment & Other

The Leopold matrix is also useful to present the results of an appraisal. This is achieved by incorporating guidelines to characterise impacts in terms of magnitude and importance on a common 1-10 scale, where 1 is least magnitude or importance, and 10 the greatest. The magnitude of an impact is taken to bc an expression of its scale, e.g. the geographical area ofthe impact. The score for importance may, however, be different. Importance refers to the significance of' the impact. If a visual impact were to occur in an area of poor landscape quality then a score of only 2 or 3 may be awarded instead of 8 or 9 in an area of high quality.

Leopold *et ul.* recommend that scores for magnitude and importance should be included in the appropriate cell of the matrix representing a likely impact. The cell should be bisected by a diagonal line with the score for magnitude being placed in the top left-hand corner. The score for importance is placed in the bottom right-hand corner. These scores can be accompanied by a plus sign (+) to indicate a beneficial effect. A completed matrix would consist of a number of cells containing two numerical scores and, in some cases, a plus sign.

The usefulness of a Leopold-type matrix has been demonstrated in the assessment of exploratory drilling sites in Antartica (Parker and Howard, 1977). Matrices were used to assess the impacts of drilling at alternative sites, and to identify the location where drilling could take place with minimum environmental disturbance. Twenty matrices were prepared to identify the 'best' site for drilling.

Despite the demonstration of the usefulness of Leopold-type matrices, they pose serious theoretical and practical limitations. Those assessing a project have to structure their thoughts by mentally linking a target component and a development activity in order to ascertain whether an impact is likely. It is then necessary to consider the next component in a similar manner. Thus, impacts are identified by a series of discrete, two-way linkages between activities and components of the environment, thereby indicating the likely direct impacts.

Lohani and Thanh (1977) modified the Leopold matrix by introducing priority values to environmental impacts. The impact of developmental project is then computed as summation of importance, magnitude and priority values for various impacted parameters as illustrated in Fig. 2.8.

2.6.2.3 Networks

The first EIA network was developed by Sorensen in 1971 to aid planners reconcile conflicting land-uses in a section of Californian coastal zone (Bisset, 1984).

Network approaches expand the concept of a matrix by introducing a cause-condition-effect relation that allows identify-cation of cumulative or indirect effects not adequately explained through simple cause-effects sequences represented by matrices. Networks, thus, recognize that a project action may trigger a series of impacts.

Network methods are based on an explicit recognition that the consequences of an action can have many ramifications for environmental and social systems. Networks provide an integrative

Fig. 2.8. Modified Leopold Matrix

Hypothetical priority Values	Proposed Action Effect on	Imigration of Labour	Dam Construction	Transmission Line	Reservoir Filling	Heavy Metal Contamination	Growth of Aquatic Weed	Relocation of Inhabitants	Total	
									Leopold Method	Lohani & Thanh (1971)
10	Health									1680
8	Spawning of fish									608
7	Archeological Artifacts									616
6	Tourism									504
5	Downstream water pollution									565
4	Social and Economic Aspects									224
3	Forestry									24
9	Fishery									180
1	Navigation									30
2	Aquatic Plants									72
Total	Leopold Method									
	Lohani & Thanh (1977)	64	103	42	286	67	61	56		

framework for impact identification and assessment which recognizes the nature of complex web of interrelationships which characterize both environmental and social systems, and the connections between them.

Network methods are not useful in impact prediction or assessment, and rely heavily on the expertise and local knowledge of persons involved in EIA. Further, for a complex project such as industrial development, the cause-condition-effect network can be unwieldy visual display.

A typical network for a paper mill EIA is presented in Figure 2.9 (NEERI, 1989).

2.6.2.4 Overlays

Overlay mappings have a long history of use in environmental planning being ideally suited for the consideration of spatial aspects. Their use in impact analysis pre-dates NEPA (McHarg, 1968). The technique relies on a set of maps of environmental characteristics (physical, biological, socioeconomic, etc.) of a project area. These maps are overlaid to produce a composite characterization of the regional environment. Impacts are identified by noting the impacted environmental characteristics within the project boundaries.

The simple method of visually representing individual mpacts and combinations of impacts has a number of advantages. The results from application of the overlay methodology are easily understood. Most importantly, it is an excellent method for showing spatial distribution of impacts. With this information, it is relatively easy to relate individual impacts and the total aggregate impact of a project . The method also allows the determination of distribution of beueficial and adverse impacts.

There are a number of disadvantages to manual overlays. Interpretation of more than twelve overlays at one time is often difficult. This means that only limited number of impacts can be considered as each impact requires separate transparency. This constraint can be overcome by the use of computers. The computerised cwerlays have been used for assessing a number of developments (Doorley and Kewkirk, 1976).

Experience with the overlay methodology indicates that it is most useful in assessing alternative routes for linear developments, such as pipelines, highways and transmission lines. Composite displays enable the impacts of routes to be assessed and also can

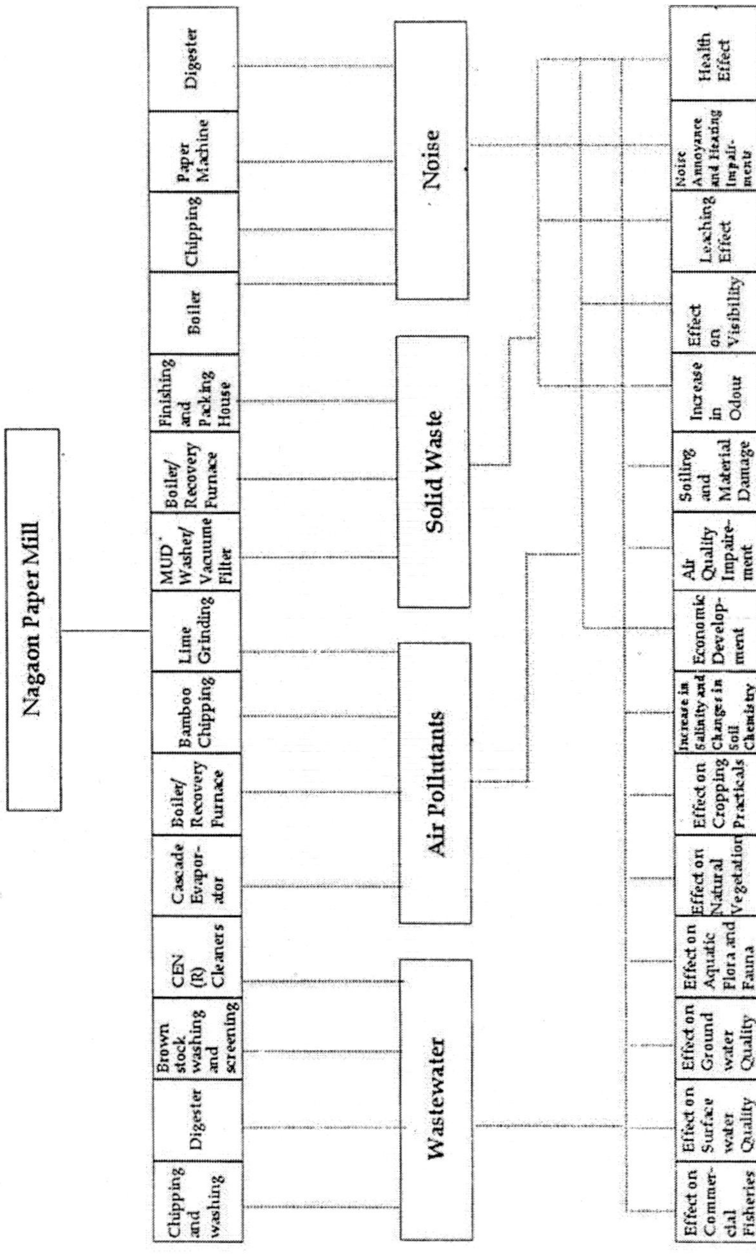

Fig. 2.9. Impact Identification Network

Contd...

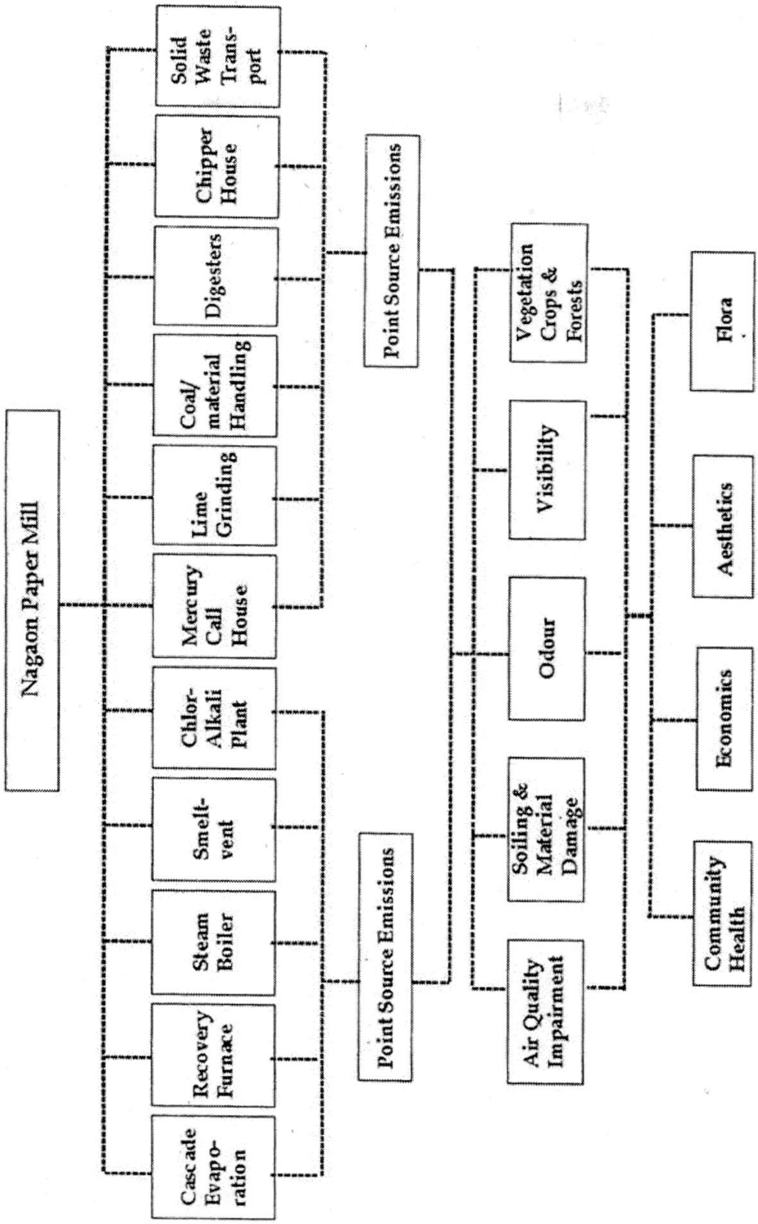
Nagaon Paper Mill
Cascade Evapo-ration
Recovery Furnace
Steam Boiler
Smelt-vent
Chlor-Alkali Plant
Mercury Call House
Lime Grinding
Coal/ material Handling
Digesters
Chipper House
Solid Waste Trans-port
Point Source Emissions
Point Source Emissions
Air Quality Impairment
Soiling & Material Damage
Odour
Visibility
Vegetation Crops & Forests
Community Health
Economics
Aesthetics
Flora

Fig. 2.9 contd...

show routes which will have least environmental impact. Therefore, overlays are very useful 'search' mechanisms.

2.6.2.5 Adaptive Environmental Assessment and Management (AEAM)

Systems analysts have evolved an approach to environmental impact assessment and management referred to as Adaptive Environment Assessment and Management (AEAM) (Holling, 1978) which combines various simulation models to predict impacts. The AEAM approach involves small interdisciplinary teams interacting through modelling workshops over relatively short period tó predict impacts and -- evaluate alternatives including management measures. The models, however, will only be as accurate and comprehensive as the data available.

2.6.2.6 Cost-Benefit Analysis (CBA)

Recognizing the fact that the traditional approaches towards EIA do not lend themselves to the integration of environmental concerns with the overall planning process, UNEP proposed an approach dealing with the evaluation of environmental impacts in monetary units to facilitate easy comparison with other costs and benefits of the project (UNEP, 1980).

The CBA is not a complete EIA methodology but comes into the EIA process only after the identification and prediction steps have been performed. Ths approach therefore, must be used with other EIA methodologies.

The CBA approach differs from the traditional EIA in that a separate document is not prepared as an EIS. Instead, environmental personnel work alongside economists and engineers throughout the entire planning process. Duplication of efforts is thus eliminated and costs of feasibility studies are reduced. Most importantly, the decision maker, when reviewing the project, would not view environmental considerations as separate or auxiliary to economic development. Although developing countries may be far from putting t h s approach into practice, it is a goal to work towards for those concerned with balancing environment and development.

2.7 COMPUTER AIDED EIA

Use of computer in EIA, with a few exceptions, has been restricted to development of software for storage and retrieval of environmental data, and software for predictive modelling.

Urban et *al.* (1975) developed a computer aided assessment system using matrix approach to identify potential environmental impacts associated with army activities in the areas of training, research and development, construction, real-estate acquisition, and industrial activities.

Krauskopf and Bunde (1972) developed a computer modeling process employing overlay technique via computer mapping for highway-route alternatives. Data on environment characteristics are collected and stored in the computer on a grid system of 1 km square cells. The programme identifies the route of least impact.

Luhar and Khanna (1988) developed a methodology for computer aided rapid EIA. The methodology consists of a predefined set of environmental predictors for baseline data collection, uses RAM model of USEPA for prediction of air quality, simple diffusion equations for predicting thermal pollution in water, and appropriate models for estimation of noise levels. The methodology uses indicators such as importance value and census,index for biological environment, and index of qualitative variation and dependency ratio for socio-economic environment. Impacts are evaluated using BEES.

The research in computer aided EIA, so far, lacks the use of comprehensive methodology for identification of impacts, identification and use of appropriate prediction models, and evaluation of impacts using predictors relevant to the type of project.

2.8 IMPACT QUANTIFICATION TECHNIQUES

The step in EIA which requires the greatest degree of scientific application is the prediction of environmental impacts. This step involves projecting the baseline environmental setting into the future, with and without the project, and then performing necessary calculations for predicting real impacts of the proposed development. The mathematical models for prediction of impacts on air, noise and water environment are well established. Indices need to be used for assessing .biological and socio-economic enviro- nment. The impact is measured as change in these indices. Indices also can be used for aggregation and communication of Impacts.

2.9 MATHEMATICAL MODELS FOR EIA

2.9.1 Concept

Models are simplified representations of reality, and can be subdivided into three main classes :

- a scaled-down copy of physical object such as building
- a mathematical representation of a physical or biological process, eg. of the spread of pollution from a stack or the movement of a weather disturbance across a region
- an exploratory representation of complex relationships amongst physical, biological, and socio-economic factors or indicators, some quantitative others qualitative.

Mathemtical models can further be classified as :

Steady state models - where the system does not vary with time. The inputs to the model are fixed and consequently the system eventually reaches some equillibrium condition. Steady state models are simplistic and have small data requirements.

Dynamic models - where the system varyies with time due to changing conditions or due to chaging inputs. Dynamic models are more difficult to formulate and require larger amounts of data. They also require larger computing time.

Deterministic models - where there is a fixed relationship between the input and output. All the relationships in such a model are rigidly determined, and the input and output are similarly assumed to be error free.

Stochastic lnodels - which allow for random fluctuations in the system due to variations in the input parameters, variations in the state of the system coefficients, or errors in output measurements. The variations are introduced by fixing the mean value of a paramter or coefficient together with the standard deviation. Data requirements for such a model are obviously greater than the equivalent deterministic model.

Mathematical model is the description of the behavior of a system in quantitative terms. It is approxin~ate description of reality and is by no means a substitution for observing the real behaviour.-Modelling exercise should be done with a clear end use in mind. In fact, well defined end-use will dictate the need for modelling (Modak, 1987).

Various steps involved in development and validation of mathematical models are shown in Figure 2.10.

2.9.2 Air Quality Models

Air quality models predict the ground level concentrations of pollutants due to specified emission sources. The processes which govern the fate of pollutants in the atmosphere include

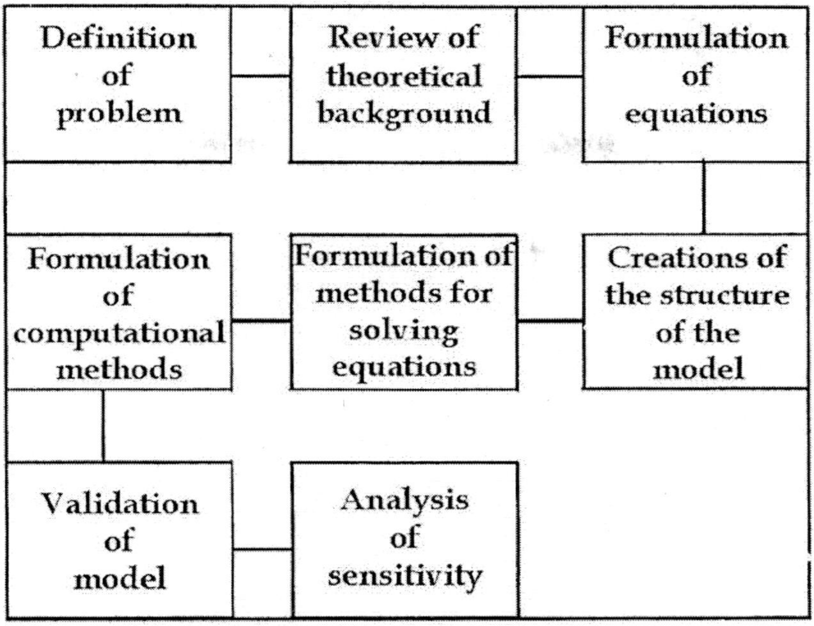

Fig. 2.10. Diagramatic representation of the steps in creating a simulation model

transportation by wind and diffusion due to turbulance. Exit velocity, emission temperature, removal of pollutants through processes such as chemical transformations, wet deposition, dry deposition, and aerodynamic effects of natural and manmade terrain features also govern the dispersion of pollutants. It is difficult to incorporate all these phenomena in a single model. Generally, a basic transport and diffusion model is formulated which is then modified to include other processes through sub-models of air chemistry, meteriological interactions etc. Air quality models can be classified based on temporal resolution, sources, receptors, and mathematical techniques.

Based on themporal resolution, air quality models can be classified as short to medium range models, and long range models. Short range (upto about 10 km) to medium-range (upto about 30 km) models are applied to areas closer to the source. Long range models examine the fate of pollutants which travel hundreds or thousands of kilometers, and generally consider the physical processes of dry and wet pollutant deposition and chemical changes. Short-range models are used mainly for hazard determination during emergencies, or for assessing impacts of new sources and their compliance with air quality standards. Long range models are used for assessing impacts of distant sources,, often in other countries. In EIA of industrial projects, short-range models are more relevant.

Mathematical techniques in air quality models include empirical, semi-empirical, and mass conservation approaches. The most commonly used air quality models are based on Gaussian Plume Model (GPM).

Gaussian model was established in 1960s bascd on thc work carried out by Pasquill and Gifford (Rau & Wootcn, 1980). The model assumes normal distributions of pollutants along verticala nd horizontal perpendicular to the direction of wind. It pclrmits assessments of continuous or instantaneous release of pollutant, with or without a linear reaction rate or decay. GPM is a semi-empiracal and semi-theoretical in nature.

The extentions of the basic Gaussian model include the following features:

- pollution reflection at the ground and at the inversion lid
- introduction of different sources (point, line, area)
- linear reaction or decay rate, washout by rain, settling of particles, uptake by vegetation or wntcbr
- topographic effects
- lake or sea breeze
- temper01 and spacial variations in mctcriological conditions

The values of diffusion coefficients in horizontal and vertical planes, sy and 'sz, are critical to the application of' GPM. Thc variation of sy and sz with distance from thc sourccl is a function of turbulance intensity which in turn is dctc.rmincd by the atmospheric stability. Computation of sy and sz by somcb commonly used methods is presented in Table 2.11 (Patil, 1984). There is substantial uncertainty and ambiguity in various methods of stability classification and diffusion coefficients determination. Accordingly, the calculated ground level concentrations vary considerably with each scheme of computation.

The initial vertical motion of a plume, known as plume rise, is an important parameter in the application of GPM. Plume rise is due to discharge momentum and thermal forces. Many divergent formulae have been develo~edto predict ~lumeri se values. Select plume rise formulae and their impact on plume rise computations are summarised in Tables 2.12 and 2.13 (Shirvaikar, 1980).

Table 2.11. Evaluation of σy and σz by some commonly used methods

Name	Stability classes	Parameters for stability classification	Ranges tested	Release and terrain condition of applicability etc.	Determination of σy/σz
Pasquill (1974)	A (extremely unstable) to F (moderately stable) (six classes)	Wind speed at 10 m and insolation during day cloud cover during night ΔT/ΔZ between 10 m and 60 m σo Net radiation	0.8 km	Ground level smoo th terrain Z=3 cm plume spread parameters given graphically for each stability category	Gra phically for each stability class as function of downwind direction
ASME (1968)	Very unstable Unstable Neutral Stable (4 classes)	Original BNL : Gustiness as shown by wind direction trace ASME : Specifies stability through ΔT/ΔZ ΔT/ΔZ for very unstable class not defined	50 km	Elevated source : Terrain not specified plume spread parameters given by power laws	Power laws
McElory (1969)	Very unstable	RiB or σo	26 km	Rough urban terrain (Z= 1m) low level source plume spread parameters given by power laws	Power laws

Table 2.12. Some widely used plume rise formulae

Name	Neutral and Uns table Atmosphere h =	Stable Atmo shpere h =	Rema rks
Briggs 1969	$1.6\,F^{1/3}(10H_s)^{2/3}$ for x > 10 hs	$2.9\left(\frac{f}{US}\right)^{1/3}$ for x> 2.4 $U/S^{1/2}$	$F = g\,QH/\pi\,Cp\,T$ $S = (g/T)\,\partial\theta$
Lucas 1967	(12.4 + 0 .09 HS) $\frac{QH^{1/4}}{U}$	Same as in Neutral and Unstable	
ISI 1978	0.84 x 4 (Lucas)	Same as in Neutral and Uns table	
ASME 1968	$150\left(\frac{F}{U^3}\right)$	$2\left(\frac{F}{UG}\right)^{0.33}$	$F = gUs\left(\frac{D}{2}\right)^2 x\,\frac{T_s - T_a}{T_a}$ $G = (g/\theta)\,\Delta\theta/\Delta_2$
Hollad 1953	$\frac{1.5\,V_s\,D + 4 \times 10^{-5}\,QH}{U}$	Same as in Neutral and Uns table	

QH = Heat emission of effluent (cal/s) D = Stack diameter (Internal) (m) Vs = discharge velocity (m/s) Q = Potential temperature of ambient air (°C) U = Wind speed (m/s) Ts = Stack gas temperature °K p = atmospheric pressure (mb) Ta = Ambient air temperature °K Hs physical stack height, m

Table 2.13. Comparison of Plume Rise Estimates (hs = 100 m, D = 5 m, Vs = 15 m/s, Ts = 150°C)

Formula	Plume Rise m	
	Neutral, Unstable*	Stable
Briggs	149	98
Lucas	160	258
ISI	134	217
ASME	169	76
Holland	58	99

*Neutral and unstable, Ta = 27°C, U_{10} = 5 m/s
Stable, Ta = 17°C, T_{10} = 2 m/s $\Delta T/\Delta Z$ = 4°C / 100 m

The software packages available from United States Environmental Protection Agency (USEPA), which are based on GPM, under the acronym UNAMAP (User's Network for Applied

Modelling of Air Pollution) are summarized in Table 2.14. In 1970s, UNAMAP models required mainframe computers. In late 1980s, Trinity consultants translated a number of UNAMAP models for IBM PC. In some cases, Trinity also developed interactive data entry programmes and graphics (Reiss, 1985).

2.9.3 Water Quality Models

Rivers have traditionally been used for the disposal of domestic and industrial wastewaters. This results in undesirable impacts on aquatic flora and fauna. The majority of these changes are brought about by the discharge of organic matter (BOD) resulting in lowering in the concentration of dissolved oxygen (DO) in the receiving water. Pollution of rivers is also caused by the discharge of toxic substances, which may break down due to chemical or bacterial action (non-conservative) or may be resistant to breakdown (conservative). Other problems arise due to discharge of inorganic nutrients causing excessive algal growth. In all these situations it is important to be able to relate the rate of discharge of pollutant to resulting concentrations in the receiving water.

An important application of mathematical modelling is to obtain a continuous profile of water quality out of observed sparse data. Once the continuous profile is obtained, it is possible to relocate or rationalize the water quality monitoring networks. Continuous profile of water quality could also be obtained for different conditions where response of water quality system could be extrapolated (simulation). Surface water quality models have been used over several decades to determine waste treatment levels and effluent standards so that the receiving streams would meet the water quality criteria for their best designated usages.

The assumptions pertaining to the mixing of pollutants in river system dictate the spacial dimensionality of the model. Although all real physical systems are three-dimensional, sufficient accuracy may be obtained in many river systems by modelling only one or two dimensions (Beck, 1987). One-dimensional models assume complete mixing in the vertical and lateral directions. Two dimensional models may assume either uniform lateral mixing, as in stratified estuaries, or vertical mixing as in relatively shallow and wide rivers. In view of availability and wider acceptibility of 1-D water quality models, the present work reviews only 1-D water quality models.

Various water quality models have been developed over the years, begining with the classic work on BODIDO models by Streeter and Phelps in the 1920s (Beck, 1987). This laid the basis

Table 2.14. Overview of Air Quality Models in UNAMAP

Model	Dispersion Equation	Source Type	Time Average	Application	Input Information	Special Features/ remarks	Out-Put
PTPLU (POINT PLUME)	Modified Gaussian Plume Equation	Single Point Source	One-hour	1. Unsuitable for complex terrain 2. No consideration of building downwash 3. Non-reactive pollutants	1. meterological data: -Amebient air temperature - Wind direction - Wind speed -Anemometer height 2. Source data -Emission information	1. Program provide an approach to estimate (i) maximum ground level concentration for comparison with ambient air quality standards and (ii) the distance to maxima 2. Program provides three options - - Stack down wash - Gradual plume rise - Buoyancy-induced dispersion 3. Brigg's plume rise equation consideration of Buoyoncy or momentum dominated plume 4. The model is applicable within 10 kms of the source. 5. Program calculates concentrations for both wind speeds constant with height and wind speeds extrapolated to stack top 6. Iterative search technique is used to find the maximum conc. and distance to the maximum conc.	1. Tables of maximum ground level concentration and their distance from the point source for bo th wind speeds constant with height and wind speeds extra polated to stack top.

PTMTP (POINT MULT-IPLE POINT)	Gaussian Plume Model	Multiple Point Sources up to 25	Hourly Av. (Number of hours up to 24 can be estimated and average conc. over this period is calcula-ted	1. Flat or gently rolling terrain 2. No aerodynamic down wash is considered 3. No wind shear in director or speed or speed is considered	1. Meterological data – wind direction – wind speed – Stability class – Mixing height – Ambient air temp 2. Source data – Emission information – Coordinates 3. Receptor data – Coordinates	1. Brigg's plume rise equation 2. P.G. dispersion parameters 3. Program produces hourly concentrations at (up to 30) receptors whose locations are specified from (up to 25) point sources 4. Wind speed and direction apply from the shortest to the tallest plume height	1. Print the information on the source receptors and inpu t meterology 2. Partial concentr-ation i.e. concentra-tion at a particular receptor because of any specific source 3. Total concentration i.e. concentration at a specified receptor due all sources.
PTDIS (POINT DISTA NCE)	Gaussian plume Model	Single Point Source	Short-term	1. Flat or gently rolling terrain 2. No aerodynamic down wash is considered	1. meterological data : – wind direction – wind speed – Stability class – mixing height – Ambient temp. 2. R eceptor data 3. Source data – Emission information	1. Program is useful in determining GLC and its variation with distance (Distance specified by user) 2. Maximum number of distances for which concentrations can be estimated is 50 3. P.G. dispersion parameters 4. Brigg's plume rise equation	1. Out put consists of listing of the source conditions and a table giving distances effec-tive stack height and conc. for each distance 2. if isopleth option is used an additional table giving the half width and a ratio of the half-width of isopleth to the half-width of sector for each isopleth value for each distance.

						5. Programm option allows the user to enter a value of concentration to be used for the determination of half-width of isopleths. For each distant if conc. exceeds the isopleth will be determined. Also this half-width of a sector of given in terms of degrees. (Maximum number of isopleths to be considered is 8.)	
HIWAY-2	Numericl Intergra-tion along the line-source of a simple Gaussion point source plume	Highway emissi-ons are consid-ered as a series of finite line source	Hourly concen-tration	1. Uncomplicated terrain – At-grade high ways – Cu-section highways 2. Non reactive pollu tants	1. Meterological data: – wind speed – wind direction – Mixing height – Atmospheric stability 2. Highway data – width – length – width of central strip – coordinates of highway and line source 3. Receptors data – coordinates – elevation 4. Traffic volume – Emission factor	1. Most applicable for ground level sources and receptors 2. Estimate of wind speed "u" at approxi-mately 2 mts height above ground is suitable. 3. Total vertical dispersion is composed of dispersion due of ambient turbulence plus the initial vertical	Hourly concentration of non-reactive pollu-tants downwind of roadways

							dispersion due to the turbulence generated by the vehicles. 4. Total horizontal dis persion is composed of dispersion due to ambient turbulence and initial vertical dispersion due to the turbulence generated by the vehicles. 5. Model accounts for the aerodynamic drag.	
CRSTER	Modified Gaussian Dispersi on Equation	Single source type but offers the cpability of consi-deringup to 19 sta cks simu ltaneou-sly loca-ted at a common site –(Chim-ney sour-ce only)	1-hr 3-hr. 24-hr. as well as ann ual mean con-centration	1. Applicable in unevan terain 2. Non-reactive pollu tants		For preprocessor program: 1. hourly-surface data (a) – wind direction – wind speed – Dry bulb temp. – Sk y cover – Ceiling height (b) –Mixing height 2. Source data – Emission information – Source elevation 3. Program control parameters and options 4. Receptor information	1. The model estimate concentration for each hour at a circular field of 180 receptor defined by – five downwind ring distances by user – 36 radials 2. Effect of limited mixing (trapping) is simulated using the method of multiple images 3. Brigg's plume rise equation only final rise is considered	1. Print outs of the input data 2. Printots of concentration table 3. Printots of concentration from sources

						4. P.G. dispersion parameters 5. The model pre-processor program calculate stability classification based on cloud cover ceiling height solar elevation (Turner method)	
VALLEY	Developed from	Single and Multiple sources up to 50 as	24-hour average concentration or option	Primary for 1. Rural area 2. Complex terrain option – Flat terrain – Urban areas 3. No fumigation condition considered	1. Meterological data: – wind direction – wind speed – Stability – Mixing height – Freq uency 2. Source data – (as an average per source for the period of concern) – Coordinates – Elevations 3. Control parameters – The user controls two factors 4. Geogra phic spacing of the receptors through a scalling factor	1. Area source contribution determined using total segment of each area source lying in the sector of concern is converted to an eff. point source lies upwind of the actual source 2. Only vertical dispersion coefficient considered. 3. Brigg's plume rise eq uation (with terrain treatment) 4. Program specifies fixed receptor network having 112 receptors defined by 16 radials (one for every 22.5 azimoth)	1. Stability wind data listing. 2. Receptor map- Each map contains normalized concentration at the 112 receptor which are located approximately on 16 radials (seven receptors each) at increments of 22.5 angular separation

						5. Half-life of pollutant is considered in dispersion equation	
MPTER (MULTIPLE POINT TERRAIN ADJUSTMENT)	Modified Gaussion Plume Equation	Multiple Point sources up to 250	Hourly (Average period longer than 1 hr. if selected by user)	1. Rural 2. Level or gently rolling terrain option slight terrain adjustment 3. No fumigation condition is considered	1.Meteriological data: – wind direction – wind speed – Stability class – Mixing height – Anemometer height 2. Source data – Emission information – Coordinates – Ground level elevation 3. R eceptor data – Coordinates – Ground level elevation 4. Control parameters	1. Program can estimate the resulting concentration at a maximum of 180 receptors from a maximum of 250 point sources 2. Model is a pplicable for source-receptor distances less than 10 kms. 3. Options are available for – stack downwash – Gradual plume rise – Buoyaney induced dispersion – terrain adjustment 5. P.G. dispersion parameters 5. Brigg's plume rise equation plume rise is considered buoyaney or momentom dominated	1. Concentration contribu tion that are available per hour and/or for the selected average period at each receptor from up to 25 sources 2. Concentration available hourly and/or for the selected averaging period at each receptor 3. Optimal out put of the following information: average concentration over length of record plus highest five concentration for each receptor for four averaging times. (1- 3- 8- and 24-hr) and additional averaging time selected by the user.

						6. Calculation of terrain adjustment is limited to receptors whose ground level elevation is less than the elevation of the lowest stack top.	
RAM	Modified Gaussian Plume Approach	Multiple Sources –Point – Area	1 hour – 1 day	1. Level or gently rolling terrain 2. Urban Areas 3. Non-reactive pollutants 4. No fumigation condition considered	1. Meterological data: – wind speed – wind direction – Stability class – Mixing height 2. Source data – All emission information for point and alongwith their coordinates 3. R eceptor data – Coordinates	1. Brigg's plume rise eq uation plume rise is considered buoyancy or momentum domin ated. Gradual rise is also consiered for the distance less than final plume rise. 2. Pasq uill-Gifford dispersion parameters 3. Transformation of a pollutant resulting in loss of that pollutant a pproximated by RAM 4. Contribution of con-centration from area sources is calculated using the Narrow Plume Simplification a pproach 5. Stack downwash considered.	1. Calculated air pollutant concent-ration at each receptor for hourly or longer averaging time specified by the user. 2. Contribution from area source and point sources are also given in output. 3. Contribution from specific Area or point source can be obtained at the option of the user.

						6. Determination of receptor locations downwind of significant sources where maxima are likely to occur.	
PTMAX (POINT MAXI-MUM)	Gaussian Plume Model	Single Point Source (For multiple stacks program can be a pplied to each stack individ ually)	Short term	1. Uniform flat terrain 2. No aerodynamic down wash condition 3. Non-reactive pollu tants	1. Meterological data: – wind speed – wind direction – Atmospheric stability – Ambient temp. 2. Emission data	1. Program is useful in determining both the maximum concentration and the distance to maximum conc. 2. In addition it gives that combination of stability and wind speed that causes maximum concentration 3. P.G. dispersion parameters 4. Brigg's plume rise equation 5. The stated wind velocity occurs at the top of the stack and applies for plume-rise and plume dilution.	1. The out put table consist of the maximum concentration and the distance to maximum concentration plume height for each combination of wind speed and atmospheric stability.

Table 2.15. Overview of QUAL-2E Model

PARAMETERS WHICH CAN BE SIMULATED	APPLICABILITY AND ASSUMPTIONS	TYPES OF SOURCE ALLOWED	LIMITATION OF MODEL
Maximum Nos. = 15 (in any combination) 1. Dissolved Oxygen 2. Biochemical Oxygen Demand 3. Temperature 4. Algae as chlorophyll a 5. Organic Nitrogen as N 6. Ammonia as N 7. Nitrite as N 8. Nitrate as N 9. Organic phosphorus 10. Dissolved phosophorus as P 11. Coliforms 12. Arbitrary Nonconservative constituents 13. Three conservative constituents	1. Dendritic (branching) streams that are well mixed. 2. Steady state or a dynamic model 3. The major transport mechanisms advection and dispersion are assumed to be one-dimensional (along the main flow of river) 4. The stream flow and waste loads are constant during the period of simulation.	1. Multiple waste discharge 2. Multiple withdrawls 3. Tribu tory inflows 4.Incremental inflow 5.Incremental ou tflow and flow augmentation	1. No. of reaches a maximum of 50 2. No. of element – No more than 20 per reach or a to tal of 500 3. No. of Head water elements - a maximum of 10 4. No. of junction element - a maximum of 9 5. No. of inpu ts and withdrawl – a maximum of 50

for modelling the biochemical kinetics of organic matter breakdown. Subsequent research has concentrated on the hydrodynamic aspects-advection and diffusion-alongwith stochastic and statistical models, and refinement of kinetic models.

Streeter & Phelps carried out work on impact of point sources of pollution and natural purification processes in the Ohio River. The method used simplistic approach of taking biochemical oxygen demand (BOD) loads as the major depleter of the dissolved oxygen (DO) and reaeration (transfer of atmospheric oxygen from air to water at the air-water interface) of the river as the source of oxygen. The saturation deficit of the dissolved oxygen is considered as the driving factor for the re-aeration phenomenon. The model used steady state condition, in which the rate of dissolved oxygen depletion was assumed to be equivalent to the rate of re-aeration.

'Most of the water quality models in actual use today are extensions of two fairly simple equations proposed by Streeter and Phelps (Loucks, 1981 and Thomann, 1982). Often used with these BOD-DO models are other first-order exponential decay, dilution, and sedimentation models for conservative and non-conservative substances. Relatively recently complex multi-parameter water quality models have been proposed and applied to predict more accurately the physical, chemical, and biological interactions of many constituents and organisms found in natural water bodies. These multi-constituent water quality simulation models generally require more data and computer time but they also provide detailed and comprehensive information on the quantity and quality of water resulting from various water and land management policies.

In the past two decades there has been an increasing emphasis on the effects of various constituents, specially nutrients, on the aquatic ecosystem, i.e. on the production of phytoplankton, zooplankton, fish and other trophic levels within the natural water bodies. Typical of many of the operational aquatic ecosystem simulation models are QUAL series of models.

F.D. Masch and A'ssociates and the Texas Water Development Board (TWAD) developed the QUAL-1 model in 1971 (TWAB, 1971). In 1985, the Centre for Water Quality Modelling of USEPA (in Athens, Georgia), in cooperation with TuRs University, modified the previous QUAL-1E model and named it as QUAL-2E UNCAS (Brown and Barnwell, 1987). Various stages of development from QUALl to QUAL2E UNCAS are summarized by Ray, 1990.

The basic equation solved by QUAL-I and QUAL-2E is the

one-dimensional advection-dispersion mass transport equation, which is numerically integrated over space and time for each water quality constituent.

Both QUAL-I and QUAL-BE require a converging stream network of computational elements and reaches with headwaters. waste discharge, water intake points, and tributories joining thr main stream. Salient features of QUALBE are presented in Table 2.15.

Temporal/Overall Model for Catchment (TOMCAT), developed by the Thames Water Authorities, United Kingdom, is another versatile model for 1-dimensional modelling of rivers. The model takes into account the diurnal variations in the wastewater and river flows with the help of Monte Carlo Simulation. The processes simulated by TOMCAT include thermal transfer, natural purification comprising carbonaceous organics oxidation, oxidation of ammoniacal nitrogen (nitrification), and oxygen balance considering these processes as sinks, and reaeration as the source of dissolved oxygen (Joshi, 1993).

Recently, Modak et al. (1989) developed 1-D water quality simulation and waste load allocation model, acronymed STREAMI. The model incorporates nitrogenous BOD pointload and non-point carbonaceous BOD in the classic Streeter-Phelps equation. The major contribution of the model relates to the use of random search technique for calibration of model paramters based on the field data.

2.9.4 Noise Prediction Models

The term noise is defined as unwanted and undesirable sound. Mathematical basis for noise prediction (sound wave divergence) is well understood (Rau and Wooten, 1980). Three basic elements involved in the study of noise predictions are the source, the receptor, and the transmisssion path. The sources of noise may be categorised as mobile and stationary sources; point, line and area sources; continuous, intermittent and implusive sources etc.

In open space sound level diminishes as the distance increases due to the geometric dispersion of the sound power. The transmission path from sourccl to rt?ceptor is characterised by the properties of the medium i.c. air, and the presence of scatterers, viz. walls, trees, vegetation etc. Noise is attenuated during transmission due to atmospheric absorption, scattered by natural and artificial barriers, and absorbed due to proynd .cover. The sound energy reaching a receptor may be due to a number of sources of different characteristics. The net energy. reaching a receptor from a number of sources may be computed by logarithmic addition of the noise level at the receptor due to all the source independently.

Generation of traffic noise varies according to volume of traffic, types of vehicles comprising the stream and their modes of operation. Once the noise is generated the resulting noise field depends upon propagation conditions such as road alignment, geometry, land topography, screening by obstacles and reflection from surfaces. Mathematical description of these processes and the resulting development of prediction techniques are used widely for assessment of traffic noise, planning and design of roads and adjoining landuse.

Federal Highway Administration (FHWA)of USA developed a model increase in noise levels due to tr&c (FHWA, 1972). The model computes hourly equivalent noise level increase due to automobiles, medium and heavy trucks separately then adds these levels logarithemically. The model takes into account vehicle type, speed,number of vehicles passing per hour, perpendicular distance from the center line of the traffic lane to receptor, absorption coefficient of ground and vegetation.

Impact on noise levels due to industrial point sources is fairly easy to compute. However, in complex situations, and consideration of barriers require use of computers. Commercial software packages for noise level impact predictions due to point sources are not widely available.

2.10 INDICATORS OF BIOLOGICAL AND SOCIO-ECONOMIC ENVIRONMENT

The complexity and variability of biological and socioeconomic systems make precise quantitative predictions of impact of industrial development impossible (Rau and Wooten, 1981). Hence, impact on biological and socio-economic environment is quantified in terms of appropriate indicators.

The assessment of potential environmental impacts warrants selection of a reference point from which biotic and abiotic changes can be measured. The reference point is usually an intact ecosystem which is compared with its affected or stressed counter-part. Indicators employed for terrestrial and aquatic ecosystem assessment alongwith the measurement techniques are summarised in Tables 2.16 and 2.17, respectively

Table 2.16. Indicators of Biological Environment : Aquatic Ecosystem

Parameter	Analysis Method	Significance of Parameters
Species Diversity (Zooplankton)	Sedwick Method Shannon Weaver Index d $d = \left(\frac{ni}{n}\right) \log 2\,(ni/n)$ ni = number of individuals of each individual species in a sample n = total number of individuals of all species in the sample	Species diversity shows the health of an ecosystem. Low species diversity is an indication of eutrophic stage.
Species diversity (Phytoplankton)	Calculation V = t x d x W x L V = volume of the sample observed (mm3) t = number of observed transects d = depth of the cell cavity (mm) W = width of the microscope field (mm) L = length of the cell cavity (mm) Organism = n(V/u) 1/C x 1000 where, n = total no. of individuals in observed transects V = volume of the sample in counting cell (mm) u = volume of observed transects (mm) Original volume of sample (ml) C = concentration factor = Volume of sample conc. (ml) Results = express as organism/L	Like zooplankton
Similarity & Pielou Indices Zoo. & Phyto.	Similarity Index = $\frac{2C}{A+B}$ A = no. of species in x sampling station B = no. of species in y sampling station C = no. of species in common to both sampl. stations Pielous Index = E = $\frac{d}{\log 2\,S}$ d = Shannon Wiever Index S = number of species in the community.	1. Similarity index analyses similarity between two aquatic ecosystems. 2. Pielou Index helps in identifying stresses on ecosystem by comparing species available in ecosystem & polluted ecosystem
Benthic fauna (microzoobenthos)	Sieve analysis to collect micro and macro inverberates. Chemical analysis to float benthic fauna on the surface. Count the number of individuals of each species.	

Commercial fisheries	Fish catch/Ha/Year -fish catch/Ha/Year fr different species of fishes from lotic & lantic habitat - average value of fish/tonnes - no. of fisherman involved.	Shows the productivity of fish & its demand in the market Analyses the harvest potential (low, high, very high).
Rare and Endangered species	List of fish available at lotic & lantic habitat is collected. The list is then compared with the Rare & Endangered species list for following five categories: Rare/ Threatened/ endangered/endemic/ entict.	Proper protection may be given to those species which are rare & endangered Identified list may help to take decision whether these plants need special at tention for growth & development.

Table 2.17. Indicators of Biological Environment : Terrestrial Ecosystem

Parameter	Analysis method	Significance of parameters
Dominance, Census and Similarity Indices	Dominance Index = Count of most dominant species/total count Census Index = no. of individuals of each species recorded/no. of kilometers covered Similarity Index = 2C/ (A+B) where A = no. of species in A station B = no. of species in B station C = no. of species common to A&B	Dominance index indicates distribution of various species Similarity index compares two sampling locations
Pest species	Weighted distribution of pest species	Helps to identify the expected impact on the productivity of any cropland. The cropland can be managed properly based on the information available on pest species
Rare and Endangered Animal/Birds species	Classification of rare and endangered animal and bird species	Proper protection may be given to rare and endangered animal and bird species if such species occurs in study area. Indentified list may help in taking decision if these species require special attention for growth & development.
Browsers & Grazers	Nos. per hectare	This parameter help to analyse carrying capacity of a grassland. If browsers & grazers are more in the grassland then the grassland can not support them so it may browse on agriculture field or encroach nearby forests.

		This parameter helps to identify the status of browsers & grazers in the ecosystem.
Terrestrial Natural vegetation	$\text{Density} = \frac{\text{No. of species A}}{\text{Area sampled}}$ Relative Density = $\frac{\text{Density of species A}}{\text{total density of all species}} \times 100$	To determine the density and relative densities of a particular species
Terrestrial Natural Vegetation	Frequency = $\frac{\text{No. of plots which species a occurs}}{\text{Total no of plots sampled}}$ Relative Frequency = $\frac{\text{Frequency value for species A}}{\text{total of frequency values of all species}} \times 100$	To determine frequency and relative frequency of a particular species in sample plots.
Terrestrial Natural Vegetation	Dominance = $\frac{\text{total cover V basal area of species A}}{\text{area sampled}}$ Relative dominance = $\frac{\text{dominance for species A}}{\text{total dominance for all species}} \times 100$	To determine dominance and relative dominance of a species in sample plots
Terrestrial natural Vegetation	Importance value = (Relative density + relative dominance + relative frequency) / 3	Importance value index is the indication of importance of different species. The species having lowest importance value has to be preserved and protected. Proper maintenance is required to protect those species having low importance value. If proper protection is not given to those species they may also become endangered in near.
Rare and Endangered plant species	List of species available in the study area to be collected from Forest Department & may be compared with Red Data Book for Indian Plants Prepare list of species under following categories: - rare - threatened - endangered - exlinct - endemic	The list give rare and endangered plant species available in the area
Pest species (Plant)	Weighted distribution of pest species present in the area	If density, frequency & abundance of pest species are more, then their immediate eradicated is necessary since these species may affect the normal growth & development of the ecosystem
Crop Productivity	$\text{Productivity} = \frac{\text{crop produced kgs}}{\text{Area hectares}}$	This shows the status of agroecosystems & its productivity
Cropping pattern	Percent area under different crops	This shows the types of crops grown in that area

The impacts of industrial project on socio-economic environment can be assessed in terms of Quality of Life (QOL). Differnt agencies have employeed indices to measure quality of life (QOL) in terms of political, economic, social, physical, and health and ambient environmental quality issues. These indices mainly vary in terms of parameteric selection for computation of QOL. A summary of parameters considered in various computations of QOL is presented in Table 2.18.

Table 2.18. Summary of select QOL Indices

Developer	Factor List
EPA	• economic environment
	• political environment
	• Social environment
	• Physical environment
	• health
	• natural environment
Perloff	• work places
	• community neighbourhood
	• household shelter
	• transportation utilities
	• natural environment
	• spatial environment
Moss	• welfare benefits
	• general government
	• regulation and safety
	• recreation and culture
	• national defence
	• human resource development
	• housing environment and utilities
	• space
Gross and Springer	• income and group values
	• urban conditions
	• crime & delinquency
	• natural environment
Wilson	• individual status
	• economic growth

	• welfare
	• individual equality
	• state & local government
	• professionalism
	• education
	• technological change
	• living conditions
	• health
	• agriculture
Flax	• income
	• unemployment
	• poverty
	• racial equality
	• community concern
	• citizen participation
	• public order
	• educational attainment
	• social disintegration
	• housing
	• transportation
	• health
	• mental health
	• air quality

According to USEPA, QOL is defined as a function between objective conditions and subjective attitudes involving a defined area of concern (EPA, 1974). The computation of quality of life involves following steps :

Computation of objective QOL index

$$QOL_0 = \sum_{i=1}^{m} Ql_i \times w$$

QOL_o = Objective QOL index

QI_i = Objective Quality Index for i^{th} factor

W = Weight of the ith factor

m = No. of factors

- Computation of subjective QOL index

$$QOL_s = 1/P \sum_{j=1}^{p} \sum_{i=1}^{m} QIij \times w$$

QOL_s = Subjective QOL index

QI_s = Subjective Quality Index (VFC) for i^{th} factor

W = Weight of the ith factor

m = No. of factors

P = Sample Population Size

Computation of aggregate QOL index

$$QOL_s = \frac{QOL_0 + QOL_s}{2}$$

The description of select parameters used in computation of objective QOL is presented in Table 2.19.

Table 2.19. Parameters Used in computation of objective Quality of Life Factors

	Major factors	Objective Indicators
1	Economic Environment	
	Income	Per capita disposable income Median family income
	Income Distribution	Gini coefficient of income distribution
	Economic Security	Income support Wealth measures
	Work Satisfaction	Accident, productivity and turnover rates
2.	Social Sector	
	Family	Marriage and divoirce rates Illegitimate births
	Community	Social responsibility scale
	Social Stability	Upward social mobility Social disorder incident rates
	Physical Security	Violent crime rates
		Human effort directed toward the arts
		Persons participating in outdoor recration and average days per person
3.	Political Environment	
	Electoral participation	Per cent of registrants voting
	Non-electoral participation	Bloomberg & Rosenstock's "Action Score"

	Government Responsibility	Budged allocations Per capita distribution of funds
	Civil Liberties	Rights Commission Citizen review board
	Informed Constituency	Content analysis of mass media
4.	Health	
	Physical	Infant mortality rate Physicians/capita Health care facility utilization
	Mental	Persons in mental hospitals/populations per cent of patients "cured"
	Nourishment	Per capita consumption of food types Nutrients consumed per day per capita
5.	Physical Environment	
	Housing	Per cent deteriorated houses Per cent lacking plumbing Per cent overcrowded
	Transportation	Family costs Per cent budged allocated to construction and maintenance
	Public Services	Cost of gas and electricity Frequency and coverage of services
	Material Quality (both goods and services)	Product life Automobile recalls Cost and frequency of repairs
	Aesthetics	Litter; billboards Trees preserved and planted
6.	Natural Environment	
	Air Quality	People exposed to sub-standard condition Concentration of CO, NO2, SO2
	Water Quality	BOD; Coliform count Turbidity; temperature; pH
	Radiation	Amount of radioactivity in water, soil, people
	Toxicity	Lead concentrations Cases of lead poisoning
	Solid Wastes	Pounds/capita Amount recycled Frequency of collection
	Noise	Community noise reference scale (under development)

2.11 ENVIRONMENTAL INDICES

Much of the initial research on environmental indices was carried out in early 1970s to measure progress towards environmental goals and pollution control targets (Ott, 1978). These were conceived as tools to describe the level of environmental quality in a particular medium for a particular region. Interregional comparisions were an important focus of the work.

Trditionally, environmental indices have been closely linked to environmental monitoring with the intent to provide insight into environmental conditions, and as a yardstick to judge the effectiveness of environmental regulatory programmes.

There are four stages in the development of a environmental index :

- Parameter selection
- Development of value function curves/scaling technique
- Assigning parameter importance units (weights)
- Selection of appropriate aggregation function

Indicators available in literature for air and water quality are summarised in Table 2.20 and 2.21 (Ott, 1978, Lohani, 1984b, Joshi, 1984).

Table 2.20. Summary of Air Quality Indices

Index	Definition	Legend
1. Green	$I_1 = 84^* X^{0.431}$ $I_2 = 26.6^* Y^{0.576}$ $I_3 = 0.5^* (I_1 + I_2)$	X = SO2 in ppm Y = coeff. of Haze
2. PINDEX	$\sum Qi/Si$	Q = ambient concentration S = standard
3. ORAQI	$5.7^* (\sum Qi/Si)^{1.37}$	Q = ambient concentration S = standard
4. ORAQI modified (TSP, SOx, NO2)	$39.02^* (\sum Qi/Si)^{0.967}$	Q = ambient concentration S = standard
5. MURC	$70^* COH^{.70}$	COH = Coefficient of Haze
6. EVI	$EVI = (\sum I^2)^{.5}$ $I = ((A8/S8)2 + (A1/S1)^2)^{.5}$	A8 = sum of observed 8-hr concentration exceedings standard S8 = 8- hr standard A1 = sum of observed 1 - hr concentration exceeding standard S1 = 1 - hr standard

8. NAQI	$\sum (IC_1)^2)^5$	IC = subindex for pollutant i
	$IC_{co} = ((C_{c1}/S_{c1})^2 + S^* (C_{c8}/S_{c8})^2)^5$ CO sub-index	S_{c1} = 1 hr standard S_{c8} = 8 hr standard C_{c1} = maximum observed 1 hr conc. C_{c8} = maximum observed 8 hr conc. S = 1 if $C_{ci} > S^{ci}$ else S = 0
	$IC_{SPM} = (C1 * C2 * C3 ... * CN/70)^{(1/N)}$ for SPM	N = number of weekly observations CN = weekly conc. in ug/cum
	$IC_{SO2} = ((C_{sa}/S_{sa})^2 + SI^* (C_{s24}/S_{s24})^2 + S2^*$ SO2 sub-index	S_{sa} = annual mean standard S_{24} = 24 hr standard S_{s3} = 3 hr standard C_{sa} = annual mean observed concentration C_{s24} = maximum observed 24 hr conc. C_{s24} = maximum observed 3 hr conc. S1 = 1 if $C_{s24} > S_{s24}$ else S1 = 0 S2 = 1 if $C_{s3} > S_{s3}$ else S2 = 0
	$I_n = C_{na}/S_{na}$ NO2 sub-index	C_{na} = annual arithmetic mean observed concentration S_{na} = annual standard
	$I_o = Co1/So1$ Photochemical oxidant index	C_{o1} = max. observed 1 hr concentration S_{o1} = annual standard 1 hr concentration

Note : Green : Air quality index developed by Green (1966)

PINDEX : Developed by Babcock (1970) based on Californian air quality standards incorporating TSP, SOX, NOx, CO, HC and OX (oxidants resulting from photochemical reaction of NOx and HC)

NAQI : National Air Quality Index by U.S. Environmental Protection Agency,

EVI : Extreme Value Index incoporating CO, Photochemical oxidants, TSP and SO2 proposed by MITRE corporation of USA

ORAQI : Oak Ridge Air Quality Index with standard values for CO, N02, OX, TSP and SO2 as 780, 400, 59, 150 and 266 uglcum, ORAQI = 100 when all concentrations are at the standards. Interpretation of ORAQI-0-20 excellent, 21-40 good, 41-60 fair, 61-80 poor, 80-100 bad, more than 100 dangerous.

Table 2.21. Summary of Water Quality Indices

		Aggregation Method	Parameter used (Weights)
1	NSF Multiplicative	$\sum I_i^{W_i}$	Turbidity (0.08), DO (0.17), pH (0.12), BOD5 (0.10), Temp (0.10), FC (0.15), Phosphorus (0.10), Nitrates (0.10), Total Solids (0.08)
2	NSF additive	$\sum I_i W_i$	Turbidity (0.08), DO (0.17), pH (0.12), BOD5 (0.10), Temp (0.10), FC (0.15), Phosphorus (0.10), Nitrates (0.10), Total Solids (0.08)
3	WQI weighted additive (Brown)	$\sum I_i W_i$	Turbidity (0.07), DO (0.18), pH (0.11), BOD5 (0.13), Temp (0.07), FC (0.18), Radioactivity (0.09), Phenol (0.10), Carb. Chlor. Ext. (0.07)
4	WQI weighted multiplicative (Brown)	$\sum I_i^{W_i}$	Turbidity (0.07), DO (0.18), pH (0.11), BOD5 (0.13), Temp (0.07), FC (0.18), Radioactivity (0.09), Phenol (0.10), Carb. Chlor. Ext. (0.07)
5	WQI Unweighted Multiplicative (Brown)	$\sum$ (Ii) 1/n	Turbidity (0.07), DO (0.18), pH (0.11), BOD5 (0.13), Temp (0.07), FC (0.18), Radioactivity (0.09), Phenol (0.10), Carb. Chlor. Ext. (0.07)
6	WQI Unweighted Additive (Brown)	$\sum$ [(1/N Ii)]	Turbidity (0.07), DO (0.18), pH (0.11), BOD5 (0.13), Temp (0.07), FC (0.18), Radioactivity (0.09), Phenol (0.10), Carb. Chlor. Ext. (0.07)
7	WQI Weighted (Solway)	$\sum \frac{1}{100}\left(\frac{1}{n} I_i W_i\right)^2$	Turbidity (0.07), DO (0.18), pH (0.11), BOD5 (0.13), Temp (0.07), FC (0.18), Radioactivity (0.09), Phenol (0.10), Carb. Chlor. Ext. (0.07)
8	WQI un weighted (Solway)	$\sum \frac{1}{n}\left(\frac{1}{100} I_i\right)^2$ 100 n	Turbidity (0.07), DO (0.18), pH (0.11), BOD5 (0.13), Temp (0.07), FC (0.18), Radioactivity (0.09), Phenol (0.10), Carb. Chlor. Ext. (0.07)

2.12 ENLARGED SCOPE OF EIA OF INDUSTRIAL PROJECTS

In EIA of industrial projects, the manufacturing process is considered as a black-box. The traditional focuss on pollution control through end-of-the-pipe (EOP) treatment which allows wasteful use of resources and then consumes further resources to solve the environmental problems in a particular medium. End-of the-pipe control technologies, more often than not, transfer pollutants from one environmental medium to another, and consume resources out of proportion to the accrued benefits (Kulkarni et al., 1991).

Future raw material and energy scenarios, the impact that the industry and its products have on the natural resource base and environmental quality, and the necessary thrust being given to industrial growth in India warrant a comprehensive strategy to deal with environmental and economic problems of the industry.

As new industries come up to meet developmental impcratives, it will be necessary to impose more stringent emission standards progressively to maintain acceptable levels of environmental quality. Hence, the options for environmental management in industry sector include, on the one hand, reactive control measures such as end-of-the-pipe (EOP) treatment technologies and media-specific regulations for waste discharges, and on the other, anticipative and preventive strategies such as adoption of cleaner technologies (CT) of production and integrated planning based on considerations of regional carrying capacity (Khanna and Kulkarni, 1992).

While in theory it is agreed that the most appropriate stage where EIA of industrial project should begin is the selection of manufacturing technology, no such case studies are available in the doccumented literature.

Incorporation of CTs in EMPs is considerably impeded by lack of information exchange between various interest groups on opportunities for pollution prevention. A meaningful data base accessible through on-line information retrieval system on availability and accessibility of CTs and their economic viability, and suitability to local environment, skills and raw materials does not exist. The available databases lack crucial information on possible savings in raw material consumption and reduction in pollution control costs through adoption of CTs (Khanna and Kulkarni, 1994).

2.13 CONCLUSION

While EIA is in practice in various countries since early 1970s, R & D endeavour in development of methodologies for EIA has been limited. This is so as many EIA practioners conceive it as an exercise involving 'technical commonsense' and involving minimum effort on quantification of impacts.

The crucial step in EIA, screening and impact identification, is aided with tools such as positive lists and checklists. This leaves substantial margin for objective decision making. Further, the checklists documented in the literature are useful in identification of primary impacts. Checklists fail to guide the project team in delineation of data requirements and identification of analysis tools.

The incorporation of environmental concerns in the process of site selection is presently enabled through extention of matrices for project level EIA. These matrices fail to include carrying capacity attributes in site selection.

The mathematical models widely used for prediction of impacts on air and water quality were developed by USEPA as regulatory tools to achieve pollution control targets. These models were developcd to handle specific situations. These packages are not user-friendly, and the output needs expert interpretation for use in EIA. Commercial packages for prediction of occupational noise levels are not yet available.

The Environmental Evaluation System (EES) developcd by Dee et al. (1973) is considered as a cornbursome methodology. The application of EES is thus rarely reported in literature. The methodology can be suitably adapted to circumvent the documented limitations through a combination of features in BEES and Leopold matrix.

The EIA project team needs to assesses subjectivity in assigning weights in EES.

The present research addresses itself to the development of software on personal computer platform to bridge these glaring gaps in the theory and practice of EIA.

REFERENCES

ADB, (1990). Environmental Guidelines for Selected Industrial and Power Development Projects, Office of the Environment, Asian Development Bank, Bangkok

Ahmad, Y. J. and Swamy G. K. (1985). Guidelines to Environmental Impact Assessment in Developing Countries, Hodder & stoughton, London

Agrawal, G. D. (1992). Environmental Impacts of Industrial Units : Assessment How and By Whom, Journal of Industrial Pollution Control, Vol. 8, No. 1, pp. 37-43

Babcock, L.R.,Jr. (1970). A Combined Pollution Index for Measurement of Total Air Pollution, Journal of Air Pollution Control Assocition, Vol 20, No. 10, pp 653-659

Brown, L. C. and Barnell, T. O., Jr. (1987). The Enhanced Stream

Water Quality Models QUAL-2E and QUAL-2E UNCAS: Documentation and User Model, USEPA Environmental Research 'Laboratory, EPAf600-3- 87-007

Bisset, R. (1984). Selected EIA Methods : An Introduction, Paper presented at the International Training Coures on Environmental Impact Assessment and Land-use Planning, Hong-kong, 16-21 January

Beck, M.B. (1987). Interpretation of Field Data in Water Quality Modelling, paper presented in International Workshop on Water Quality Management, January 12-16, IIT, Bombay

Canter, L.W..(1977). Environmental Impact Assessment, MCGraw Mill, New York

Canter, L.W., and Hill, L.G. (1982). Handbook of Variables for Environmental Impact Assessment, Ann Arbor Science Publishers Inc, Michigon

Clark, B.D., Chapman, K., Bisset, R., Wathern, P. and Barret, M. (1981). A Manual for the Assessment of Major Developments, HPIISO, London

Commission of Europian Communities (1985). On the Assessment of ffects of Certain Public and Private Projects on the Environment, Official Journal, L175, 28.5.85

Dee, N., Baker, J.K., Drobny, N.L., Duke, K.M., Whitman, I. and Fahringer, D.C. (1973). An Environmental Evaluation System for Water Resources Planning, Water Resources Research, vol. 9, no. 3, pp. 523-535

Dooley, J. E. and Newkirk, R. E. (1976). corridor Selection Method to Minimize the Impact of an Electrical Transmission Line, Toronto, James F. MacLaren Ltd.

EPA, (1974). Studies in Environment - Vol. I1 - Quality of Life, EPA-6001 5-73-012b, Office of Reserach and Development US .Environmental Protection Agency, Washinton D.C.

FHWA, (1972). Manual for Highway Noise Prediction, Report No. OST-TSC-FHWA-72-1, Federal Highway Administration, Washington, D.C.

Green, M. H. (1966). An Air Pollution Index Based on Sulphur Dioxide and Smoke Shade, Journal of Air Pollution Control Association, Vol. 12, pp. 703-710

Garner, J.' F. and O'Riordan, T. (1982). Environmental Impact Assessment in the Context of Economic Recession, Geographical Journal, Vol. 148, No. 3, November, pp. 343-361

Go, F.C. (1990). Environmental Impact Assessment, MARC report no. 41, Monitoring Assessment Research Centre, London

Holling, C.S. (1987). Adaptive Environmental Assessment &

Management, John Willey & Sons, New York

Htun, N. (1988). The EIA Process in Asia and Pacific Region in P. Wathern ed. Environmental Impact Assessment : Theory and Practice, UNwin Hyman, London

IUCN, (1980). World Conservation Strategy - Living Resources Conservation for Sustainable Development, International Union for Conservatio~ol f Nature and Natural Resources, Gland, Switzerland

Joshi, V. (1993). Water Quality Models and their Application in EIA, paper presented in Workshop on Methodologies and Procedures in EIA of Industrial and River Valley Projects, November 22-26, NEERI, Nagpur, India

Krauskopf, Thomas, M. and Dennis, C. Bunde (1972). Evaluation of Environmental Impact Analysis : Philosophy and Methods, ed. By Robert Ditton and Thomas Goodale, University of Wisconsin

Khanna, P., Kulkarni, V.S., Joshi, V., Dutt, P.S., Agarwal, A.L. (1991). Carrying Capacity Based Developmental Planning Process, Journal of Indian Association of Environmental Management, Vol. 18, No. 1&2, pp. 12-19

Kulkarni, V.S., Saraswat, N. and Khanna, P. (1991). Development and Implementation of Cleaner Technologies of Industrial Production, Journal of Indian Association for Environmental Management, Vol. 18, NO. 1852, pp. 59-69

Khanna, P. and Kulkarni, V.S. (1994). Transfer of Environmentally Sound Technologies : Co-operation and Capacity Building, report submitted to Economic and Social Commission for Asia and Pacific, Bangkok

Khanna, P. and Kulkarni, V.S. (1992). Policy Options for Environmentally Sound Technologies for Sustainable Development, ESTA Bulletin of United Nations Council for Science, Technology and Deve!opment, Issue No. 7, Spring 1992, pp. 127-133.

Leopold, L.B., Clarke, F.E., Hanshaw, B.B. and Balsley, J.R. (1971). A Procedure for Evaluating Environmental Impact, US Geological Survey Circular 645, US Geological Survey, Washington D.C.

Lohani, B.N., and Thanh, N.C. (1977). Rural Development and its Environmental Impact Assessment in South-East Asia, Proceedings of International Copference on Rural Development Technology, An Approach, AIT, Bangkok

Loucks, D.P. (1981). Water Quality Models for River Systems in Models for Water Quality Management ed. A. K. Biswas, McGraw-Hill International Book Company, New York

Luhar, A.K., and Khanna, P. (1988). Computer Aided Rapid Environmental Impact Assessment, Environmental Impact Assessment Review, No.8, pp. 9-25

McHarg, I. (1968). A Comprehensive Route Selection Method, Highway

Research Record 246, Highway Research Board, Washington, D.C.

MEF (1985). Environmental Guidelines for Siting of Industrial Projects, Ministry of Environment and Forests, Government of India.

MEF (1990). Annual Report - 1989-90, Ministry of Environment and Forests, Government of India

MEF (1991). Annual Report - 1990-91, Ministry of Environment and Forests, Government of India

MEF (1992). Annual Report - 1991-92, Ministry of Environment and Forests, Government of India

MEF (1993). Annual Report - 1992-93, Ministry of Environment and Forests, Government of India

MEF (1994). Annual Report - 1993-94, Environmental Impact Assessment Notification : 1994 (as amended on 4.5.1994), Ministry of Environment and Forests, Government of India

Modak, P. M. (1987). Introduction to Water Quality Modelling paper presented in International Workshop on Water Quality Management, 12-16 January, IIT Bombay, India

Modak, P. M., Gelda, R. K., Clemant Prabhakar, T., Venketwaralu, S. and Gopakumar (1989). Water Quality Modelling on Personal Computers - Case of STREAM Software, Journal of Indian Water Works Association, Vol. 21, No. 1, October- Decemeber, pp 315-322

Munn, R.E. (1979). Environmental Impact Assessment : Principles of Procedures, International Council of Scientific Unions, Scientific Committee on Problems of Environment, Report -5 (SCOPE -5)

NBRO (1988). Environmental Assessment Procedure for Development Activities in Urban Areas : A Manual for Planners and Decision Makers, National Building Research Organization, Ministry of Local Government Housing and Construction, Sri Lanka

NEERI (1989). Rapid Environmental Impact Assessmnet of Nagaon Paper Project, National Environmental Engineering Research Institute, Nagpur, India

OECD (1986). Recommendations of the Council on Measures required to Facilitate the Environmental Assessment of Development Assistance Projects and Programmes, C(86126, Organization for Economic Co-operation and Development, Paris

OECD D.evelopment Assistance Committee (1992). Guidelines on Environment and Aid : Good Practices for Environmental Impact Assessment of Developmental projects, Development Co-operation Directorate, Paris

Ott W.R., Thom G.C. (1976). A critical review of air pollution index systems in United States and Canada, Journal of Air Pollution Control Association, Vol. 26, No. 5, pp. 460-470

Ott, W. R. (1978). Environmental Indices. Ann Arbor Science Publishers Inc. Michigan

Parker, B.C. and Howar, R.V. (1977). The First Environmental Monitoring and Assessment in Antartica : The Dry Valley Drilling Project, Biological Conservation, Vol. 12, No. 2, pp. 163-177

Patil, R.S. (1984). Air Quality Modelling, paper presented in WHO sponsored training programme in Air Pollution Monitoring and Management, October 1-31, IIT, Bombay

Rau, J.G., and Wotten, D.C. (1980). Environmental Impact Analysis Hand Book, Mc Graw-Hill, New York

Reiss, N. M. (1985). Microcomputer Software Review, Journal of Air Pollution Control Association, Vol 37, No. 7, pp. 769-796

Sassaman R. W. (1981). Threshold of Concern : **A** Technique for Evaluating Environmental Impacts and Amenity Values, Journal of Forestry, vol. 79, No. 2, pp. 84-86

Schaenman, P.S. (1976). Using and Impact Measurement System to Evaluate Land Development, The Urban Institute, Washington D.C.

Shirvaikar, V.V. and Kapoor, R.K. (1980). Air Pollution Models : An Assessment of Uncertainities from Users Point of View in Patel, B., ed., Management of Environment, Wiley Eastern Limited, New Delhi

Solomon, R.C., Colbert, B.K., Hansen, W.J., Richardson, S.E., Canter, L.W. and Vlachos, E.C. (1977). Water Resources Assessment Methodology (WRAM) : Impact Assessment and Alternative Evaluation, 'Technical Report Number Y-77-1, US Army Corps of Engineers, Vicksburg, Mississippi

Sondheim, M.W. (1978). A Comprehensive Methodology for Assessing Environmental Impact, Journal of Environmental Managemqnt, vol. 6, no. 1, pp. 27-42

Streeter, H. W. and Phelps, E. B. (1925). A Study of the Pollution and Natural Purification of the Ohio River, Bulletin of the US Public Health Service

Thomann, R. V. (1982). Verification of Water Quality Models, Journal of Environmental Engineering Division, American Society of Civil Engineers, vol. 108, no. EE5, pp 923-940

Tomlinson, P. (1984). Project Screening Methods paper presented in International seminar on Environmental Impact Assessment. University of Aberdeen

UNCED, (1992). Rio Declaration, United Nations Conference on Environment and Development Rio de Junerio

UNEP. (1980). Environmental Assessment Statement - A Test Model Presentation, UNEP Regional office and UN Asian and Pacific Development Institute, Bangkok

UNEP, (1980). Guidelines for Assessing Industrial Environmental Impact and Environmental Criteria for the siting of Industry, UNEP Industry and Environment Guidelines Series, Vol.1

UNEP, (1987). Environmental Law Guidelines and Principles : Environmental Impact Assessment, Narobi

UNEP, (1988). Environmental Impact Assessment : Basic Procedures for Developing Countries, Narobi

Urban, L.V. (1975). User Manual-Computer Aided Environmental Impact Analysis for Cosntruction Activities, Technical Report E-50, Construction Engineering Research Laboratory, Washington, D.C.

USAID, (1980). Environmental Design Considerations for Rural Development Projects, US Agency for International Development, Washington D.C.

Wathern, P. (1988). Introduction in Wathern, P (ed) Environmental Impact Assessment : Theory and Practice, UNwin Hyman, London

Wathcrn, P. (1988). The EL4 Directive of the Europian Community in P. Wathern ed., Environmental Impact Assessment : Theory and Practice, Unwin Hymann, London

WCED (1987). Our Common Future, World Commission on Environment and Development Oxford University Press, London

Whitman, I.L. (1971). Design of Environmental Evaluation System, Battelle Columbus Laboratories, Columbus, Ohio

Wood, C. (1994). The Environmental Assessment of Plans Programmes and Policies : A Comparative Review, Indo-British Workshop on Environmental Impact and Risk Assessment of Petrochemical Industry and Environmental Audit, January 8-10, 1994, Nagpur, India

World Bank (1987). Environment, Growth and Development, Development Committee Paper 14, The World Bank, Washington D.C.

World Bank (1989). Operational Directive 4.00, Environmental Assessment, The World Bank, Washington, D.C.

World Bank (1991). Operational Directive 4.01, Environmental Assessment, The World Bank, Washington, D.C.

CHAPTER — 3

EIA-AID : A SOFTWARE PACKAGE FOR COMPUTER AIDED EIA

3.1 PACKAGE ORGANIZATION

The present research is based on the definition of EIA given by Heer and Hagarty (1977). According to this* definition, Environmental Impact Assessment (EIA) consists in establishing quantitative values for selected parameters which indicate the quality of the environment before, during, and after the proposed development activity.

The overall framework for EIA of industrial project developed herein comprises :

- Screening of proposals to decide whether EIA studies are required, and identification of generic impacts
- Selection of site for the project based on objective evaluation of the site alternatives
- Site specific studies to evaluate impacts of the proposed project on environmental quality including
  - identification of site specific impacts
  - collection of baseline data
  - prediction of impacts
  - evaluation of impacts
  - sensitivity analysis in impact evaluation
- Design of green-belt for abatement of fugitive emissions

A computer software package, acronymed EIA-AID, has been developed in this research to assist the project team in planning and executing EIA studies. The software is a menu driven package developed for PC environment in Quick Basic 4.5, and has following modules:

- Screening of project proposals
- Ranking of site alternatives
- Identification of impacts
- Impact prediction
- Impact evaluation
- Sensitivity analysis

Green-belt design

The overall organization of the package is presented in Fig. 3.1. EIA-AID is stand-alone package and can work on PC-XT with monochrome monitor. However, to harness the best features, PC-386 with EGA monitor is desirable. Thc basis, algorithm, input data requirements, outputs, and salient features for each of these modules are discussed hereunder.

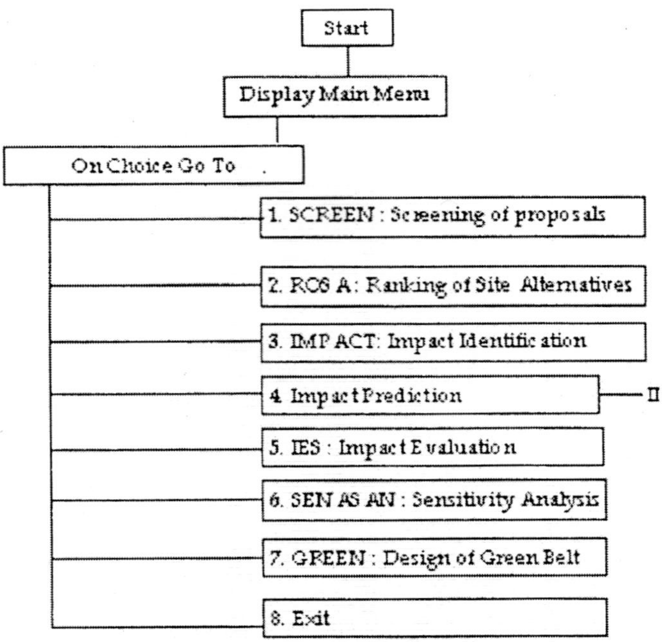

Fig. 3.1. System flow-chart for EIA-AID icontd ...)

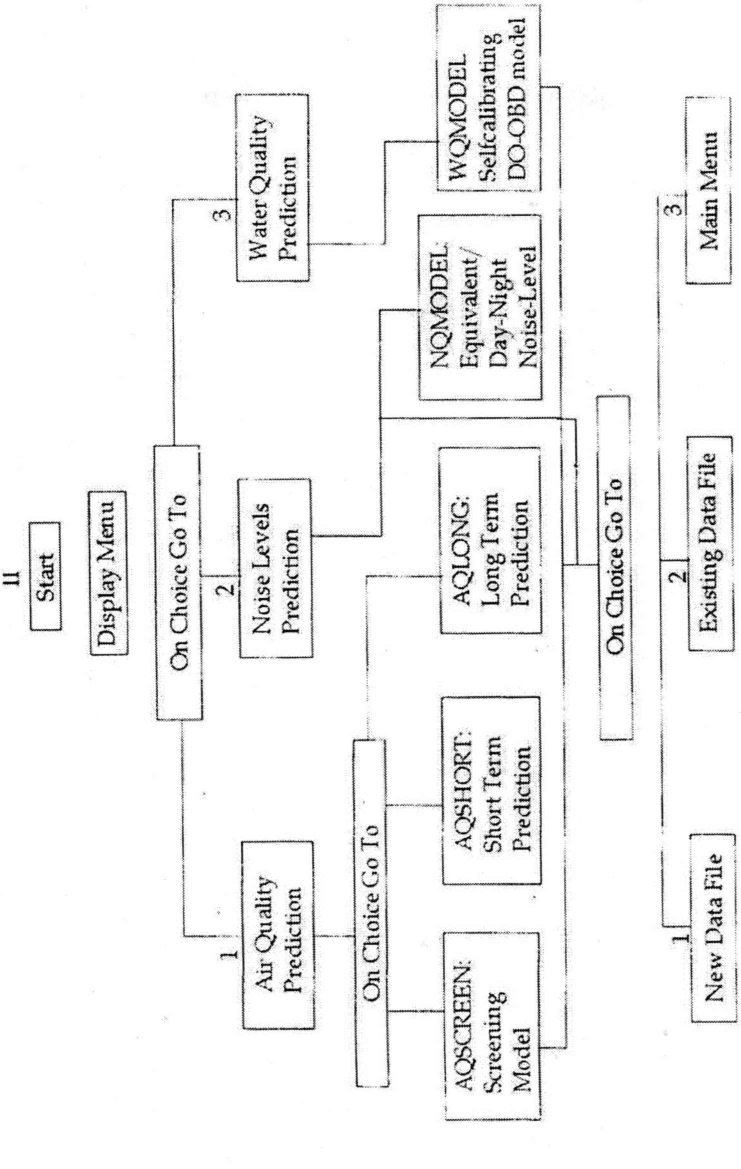

Fig. 3.1. System flow-chart for EIA-AID

3.2 SCREENING OF PROJECTS

The screening module developed in the present research is based on the 'project threshold' methodology for screening. Industrial projects have been categorised as :

Category A : Projects requiring comprehensive EIA

Category B : Projects requiring rapid EIA

Category C : Projects which need not be subjected to EIA

The Ministry of Environment and Forests in the notification of EIA of January 28, 1993 proposed the listing of projects to be appriased at the Central and State Government levels. This listing forms the basis for screening chieria proposed herein (Table 3.1).

Table 3.1. Screening criteria for EIA of Jndustrial projects

S. No.	Project type Category A	Threshold for Category B
1	Petroleum refineries including crude and product pipe line and storage of petroleum products	All
2	Chemicals fertilizers (nitrogenous and phosphatic) other than single superphos phate	All
3	Pesticides All	
4	Petrochemical complexes (both Olef-Inic and Aromatic) and petro chemical intermediates such as DMT caprola-ctum LAB etc., and Production of basic plastics such as LDPE, HPDE, PP, PVC etc.	All
5	Industrial explosives including detonating fuse, safety fuse, gun powder nitrbcellulose electric detonators	All
6	Drugs and pharmaceuticals (basic and bulk)	All
7	Exploration for oil and gas and their production, transporation and storage	All
8	Synthetic rubber All	
9	Asbestos and asbestos products	All
10	Hydrocyanic acid its derivatives	All
11	Primary metallurgical industries (such as production of iron and steel, aluminium, copper, zinc, lead and ferro alloys)	All

12	Pneumatic tyres and tubes of rubber used in buses and trucks, air crafts, jeeps, motor cars, tractors and of the road vehicles	All	
13	Chlor alkali industry	All	
14	Integrated paint complex including manufacture of resins and basic raw materials required in the manufacture of paints	All	
15	Man-made fibreslfilament yarn	All	
16	Storage batter is integrated with manufacture of oxides of lead and lead antimony alloy	All	
17.	Cement	>200 TPD	
18	Sponge iron		
	a. Coal based	>30000 TPA	upto 30000 TPA upto 500000 TPA
	b. Gas based	>500000 TPA	
19	Raw skins and hides	5000 skins per day	upto 5000 skins per day
20	Distilleries	150 Wday	upto 150 KLIday
21	Sugar	2500 TP	upto 2500 TP
22	Textiles	500 meters per day	upto 500 metres per day
23	Pulp paper and newsprint	33000 TPA	upto 33000 TPA
24	Dyes	>5 TPD	upto 5 TPD
25	Dye intermediates	>5 TPD	upto 5 TPD
26	Thermoset resis	>5000 TPA	upto 5000 TPA
27	Acids	>500 TPD	upto 500 TPD
28	Animals or vegetables fats oils and their fractions partly or wholly hydrogenated	>25 TPD	upto 25 TPD
29	Foundaries (individual)	> Rs. 2 crores of outlay	upto Rs. 2 crores of outlay

The Screening module enables selection of projects for rapid/ comprehensive EIA based on the project threshold database developed during this research, and inbuilt in the programme. The module also enables preliminary screening of site options based on siting criteria delineated by the MEF. The information provided by the user is compared with the inbuilt database to decide the project category, and site alternative acceptibility. The flowchart for the screening is presented in Fig. 3.2.

3.3 RANKING OF SITE ALTERNATIVES

Site should ideally be decided on the basis of the regional carrying capacity. There are certain criteria stipulated by MEF

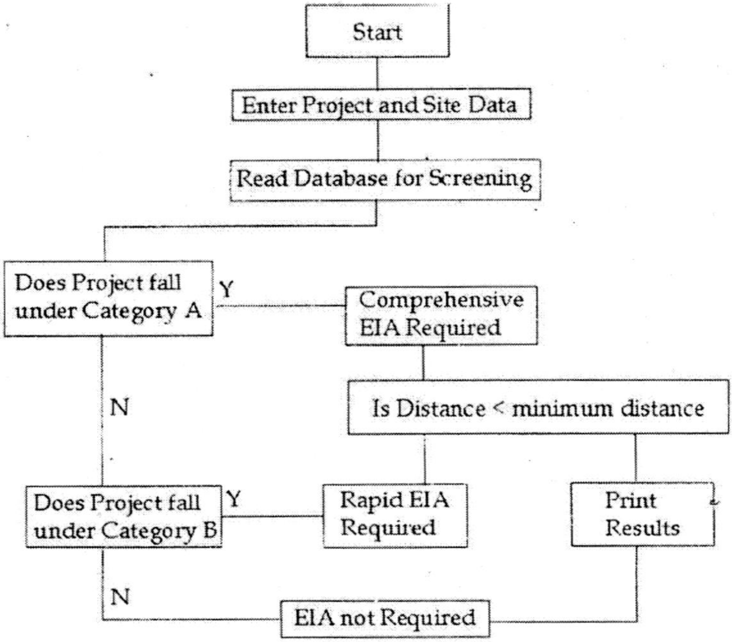

Fig. **3.2.** System flow-chart for SCREEN : Screening of Proposals

(1985) based on which a site could be directly rejected without undertaking detailed assessment of the supportive and assimilative capacities of the region.

Site selection warrants conversion of the site specific environmental attributes as also project attributes to a comparable set of units using an appropriate system of normalization. The methodology developed in the present research uses scalingweighing method for Ranking of Site Alternatives (ROSA) based on assessment of regional carrying capacity. The parameters in ROSA are divided in four categories, viz.

- Supportive capacity
- Assimilative capacity
- Human interest
- Project attributes

These categories have been further divided into 16 components which group all 73 paramters as shown in Table **3.2.** Measurement units for these parameters are delineated in Table **3.3.**

Table 3.2. Attributes for Ranking of Site Alternatives

Category	Component	Parameters
Supportive Capacity	Water	Surface water Availability Ground Water Depth Ground Water Requirement
	Power	Power Requirement Power Source
	Infrastructure	Road Transport-Highway Rail Transport-Station Port/Harbour Housing Facilities Medical Facilities Skilled Labour
	Land	Forest Land for Project Agri. Land for Project Waste Land for Project
Assimilative capacity	Baseline air quality	Suspended Particle Matter Sulphur Dioxide Nitrogen Dioxide Carbon Monoxide Hydrocarbons Photochemical Oxidants
	Baseline noise levels	Noise Level-sparce Noise Level-frequent Noise Level-continuous
	Baseline water quality	Dissolved Oxygen Biochemical Oxygen Demand pH, Temperature Faecal Coliform Total Phosphates Total Nitrayes Turbidity, Total Solids
	Baseline soil quality	Exchangable sodium % Water Extractable Metals
	Air assimilative capacity	Topography Wind Speed Wind Direction towards Receptor Inversions Mixing Depth Ventilation Coefficient
	Water assimilative capacity	Dilution Potential Flow Velocity
	Land assimilative capacity	Soil Permiability

	Ecologically sensitive areas	Distance to Historic Site Distance to Archecological site Distance to Biological Reserves Distance to National Park Distance to Sanctuaries
Human Interest	Socio-economics	Work Participation Community Health Population Density
	Aesthetics	Per capita Income
Project Attributes	Production cost	Landscape Vegetation Cover
		Transport Water Power Labour Land Construction Environmental Management Total cost
	Commissioning & operation	Land Aquisition Problem Labour Problems Government Co-operation

Table 3.3. Parameter Measurement for Ranking of Site Alternatives

Sr. No.	Parameter	Measure-ment Unit .	Measurement! Significance
	Category : Supportive capacit Component : Water		Parameter measurement is subjective. Enter on
1 .	Surface Water Availability	0 – 1	Parameter measurement is subjective. Enter on a scale of 0 to 1, where **0** correspondence to poor reliability and 1 corresponds to very high reliability of the surface water source for water supply to the project after meeting other competing demands.
2	Ground Water Depth	m.	In the case of unconfined aquifer, as the depth of water table increases with depth. Also, other competing uses suffer because of lowering of water table because of industrial water use. Further, lowering of water table may adversely affect water quality.
3	Ground Water Requirement	Demand/ yield	Ratio of water requirement of the project from ground water and yield of the aquifer during the critical period.
	Component : Power		
	1. Power Requirement	demand/ availabi-lity	Ratio of ower requirement of the project and surplus power available from the existing source /national grid after meeting the competing requirement during critical period (peak loads)

2.	Power Source	Pollution Potential	If the source of power is itself highly polluting then in turn it will lead to off-site pollution eg. coal based thermal power station. Hence assign pollution potential on the scale 0 to 1 where 0 corresponds to coal based thermal power station and 1 corresponds to renewable energy source. The measurement is qualitative and subjective.

Category : Supportive Capacity
Component : Infrastructure

1	Road Transport Highway	distance Km	Any industrial project would involve transport of raw material and finished products. In absence of transport linkages, it will warrant development/ construction of new facilities. Accordingly distance to existing nationaV state highway, railway station, port and harbour, and air port is an important consideration depending on the nature of the project.
2	Rail Transport-Station	distance Km	
3	Port/Harbour	distance Km	
4	Housing . Facilities	0-1	Parameter measurement is subjective. Enter on the scale 0 to 1 where 0 corresponds to poor housing facilities within say, 10 km radial distance of the proposed site and 1 corresponds to very good housing facilities.
5	Medical Facilities	0-1	Parameter measurement is subjective. Enter on the scale 0 to 1 where 0 corresponds to poor medical facilities within say, 10 km radial distance of the proposed site and 1 corresponds to very good medical facilities. Apart from routine facilities, this is an important aspects with respect to disaster management and emergency preparedness
6	Skilled Labour	0-1	Parameter measurement is subjective. Enter on the scale **0** to 1 where **0** corresponds **to** non-availability of local skilled labour, say within 10 km radial distance of the proposed site and 1 corresponds to very good availability of skilled labour locality. Non-availability of local skilled labour would result in migration.
7	Unskilled Labour	0-1	Parameter measurement is subjective. Enter on the scale 0 to 1 where 0 corresponds to non-availability of local unskilled labour, say within 10 km radial distance of the proposed site and 1 corresponds to very good availability of unskilled labour locally. Non-availability of local unskilled labour would result in migration.

Category : Supportive Capacity
Component : Land

1	Forest Land for Project Aquired	% of total land to be	As per guidelines issued by the Ministry of Environment and Forest no forest land can be converted into industrial area. However, for socio-political reasons such conversion if deemed necessary, the extent of forest land to be affected should be minimum.
2	Agricultural Land for Project	% of total land to be aquired	As per guidelines issued by the Ministry of Environment and Forests no prime agricultural land can be converted into industrial area. However, for socio political reasons such conversion if deemed necessary, the extent of land to be affected should be minimum.
3	WasteLand for Project aquired.	% total land to be	Any attempt to use land designated as wasteland for industry is welcome from ecology point of view.

Category: Assimilative capacity
Component : Ambient Air Quality

1.	Suspended Particulate Matter	ug/cum	Average annual 24-hrs concentration in the pre-dominant downwind direction of the Proposed facilities.
2.	Sulphur Dioxide	ug/cum	Average annual 24-hrs concentration in the pre-dominant downwind direction of the proposed facilities.
3.	Nitrogen Dioxide	ug/cum	Average annual 24-hrs concentration in the pre-dominant downwind direction of the proposed facilities.
4.	Carbon Monoxide	ppm	Average annual 8-hrs concentration in the pre-dominant downwind direction of the proposed facilities.
5.	Hydrocarbons	ppm	Average annual 8-hrs concentration in the pre-dominant downwind direction of the proposed facilities.
6.	Photochemical Oxidants	ub/cum	Average annual 8-hrs concentration in the pre-dominant downwind direction of the proposed facilities.

Category : Assimilative Capacity
Component : Ambient Water Quality

1.	Dissolved Oxygen	mg/1	Maximum deviation downstream of the waste-water discharge point of the proposed project
2.	Biochemical Oxygen Demand	mg/l	Maximum deviation downstream of the waste-water discharge point of the proposed project
3.	pH	deviation form 7.0	Maximum deviation downstream of the waste-water discharge point of the proposed project
4	Temperature	deviation 0C from ambient	Maximum deviation downstream of the waste-water discharge point of the proposed project

5	Faecal Coliform 100ml	MPNI	Maximum concentration downstream of the waste water discharge point of the proposed project
6	Total Phosphate	mg/l	Maximum concentration downstream of the waste water discharge point of the proposed project
7	Total Nitrates	mg/l	Maximum' concentration downstream of the waste water discharge point of the proposed project
8	Turbidity	JTU	Maximum concentration downstream of the waste water discharge point of the proposed project
9	Total solids mgll		Maximum concentration downstream of the waste water discharge point of the proposed project
	Category : Assimilative Capacity Component : Soil Quality		
1	Exchangeable Sodium %	ESP	If Exchangeable sodium percentage is more than 15.0 then soil becomes impermiable & sticky when wet and very hard when dry. ESP is calculated as ratio of concentration of sodium to total concentration of sodium, potassium, calcium and magnesium.
2	Water Extractable Metals	mg/1	If soil pontains water extractable metals and if project aims at disposal of wastewater on land then it may lead to pollution of ground and surface water sources.
	Category : Assimilative capacity Component Ambient Noise Level		
1	Noise level sprace	dB	Noise level in the immediae vicinity of the proposed site say within a distance of 2 km.
2	Noise level frequent	dB	Noise level in the immediae vicinity of the proposed site say within a distance of 2 km.
3	Noise level continuous	dB	Noise level in the immediae vicinity of the proposed site say within a distance of 2 km.
	Category: Assimilative Capacity Component : Air Assimilative Capacity		
1	Topography	0-1	Parameter measurement is subjective and is indicative of impact of topography on dispersion of pollutants. Enter in the scale 0 to 1 where 0 corresponds to valleyheat island type situation and 1 corresponds to flat gently rolling terrain.
2	Wind speed	m/s	Maximum ground level concentration of air pollutants is inversely proportional to the average wind speed.
3.	Wind Direction towards sensitive Receptors	%	It is preferable to locate the project downwind of the sensitive receptors to minimize adverse impacts.

4	Inversions	nos./year	Inversion contains the pollutants and hence poses risk to human health, plants animals and materials.
5	Mixing Depth .	m	Dispersion of pollutants is directly proportional to mixing depth in neutral and unstable meterological conditions. The parameter measurement should be for minimum mixing depth at the site.
6	Ventilation Corfficient	m/s	Ventilation coefficient is calculated as multiplication of mixing depth and mean wind speed through the mixing depth. It provides an indirect measurement of assimilative capacity. Dispersion of pollutants in directly proportional to ventilation coefficient in neutral and unstable meterological conditions. The parameter measurement should .be for winter.
Category : Assimilative Capacity			
Component : Soil Assimilative capacity			
1.	Soil Permiability	porosity	It is indirect measurement of soil Permiability. Lesser the permiability, chances of ground water pollution are less. Data is rcxquired for site where wastes are to be disposed.
	Category : Assimilative Capacity		
	Component : Ecology		
1.	Distance to Historic Site	Km.	All these areas are designated as ecologically sensitive areas by the Ministry of Environment and Forests. It is desirable to locate developmental activities awav from these areas. Distance are crow flying distances from the proposed project site
2.	Distance to Arch iological Site	Km	
3.	Distance to Biosphere Reserve	Km.	
4.	Distance to National Park	Km.	
5.	Distance to Sanctuaries	Km.	
	Category : Human interest		
	Component : Aesthetics		
1	Landscape	0-1	Parameter measurement is subjective. Enter on the scale ***0*** to 1 where ***0*** corresponds to unpleasant/poor landscape and 1 to a beautiful and pleasing landscape.
2	Vegetation Cover	% Pr. Area	Development of facilities.for the project may require clearing of vegetation cover at the site. Although same could be compensated to albeit to a limited extent through aforestation and development of green belt, it is desirable to limit disturbance to the natural veget ation minimum.

Category : Human interest
Component : Socio Economics

1	Work Participatino	ratio	Population in the age group 15 to 60 is considered active for which employment opportunities are to be provided. Census records provide information on the number of workers of all categories. Work participatton ratio is the ratio of working Lo active population. Ideally this should be closer to 1.0. The socio-political policies warrant that the development activities be placed in backward areas so as to prevent rural-urban migration for jobs
2.	Community Health	0-1	Parameter measurement is subjective. Enter on a scale 0 to 1 where 0 corresponds to poor comm-mtnunity health i.e. predominance of various dlseases related to water and air pollution and 1 corresponds to very good cotnmunity health.
3.	Population Density km.	Populati on per sq.	Any developmental activity would result in imigration resulting in increase in population density i.e. total population within, say. 10 km. radial distance of the site divided by the area. Population density as per 1981 census was 216 per sq. km.
4.	Per Capita Income	Rs/Yr	Per capita income is indirect measurement of the purchasing power Data is available in Census recors.

Catefory : Project Atteributes
Componet : Production Cost

1	Transport	0-1	Financial analysis of a project would always favour an alternative that would lead to maximization of (private) profits. Accordingly, from project proponents point of view the alternative are to be ranked in such a way that maximum cost site alternative would be unacceptable (score = 0) while the least cost alternative would be preferred (score =I). Score for the other site alternatives is to be linearly interpolated. In absence of data the measuremerit can be subjecttive, while the logic would remain unchanged.
2	Water	0-1	Cost of transport includes transporation of raw materials from the point of supply as also finished products to the point of distribution: per unit of products.
3.	Power	0-1	Cost of water includes total cost of water production for various requirements and may include cost of source dcveloptnent intake works. Raw water conveyance and water treatmert.
4	Labour	0-1	Cost of power includes total tariff paid/ expenditure incurred for meeting the power requirementsof the project. As the tarrif rates vary from state to state. this could be an important attribute.

5	Land	0-1	Cost of labour includes cost of skilled and unslu-Iled Iahoul as also manegorial and support staff.
6	Construction	0-1	Cost of land is the total cost of land acquisition for the project.
7			Cost of environmental management includes cost of pollution control to meet the site specific st;lndards (stipulated by the concerned State Pollution Control Board), disposal of wastes, compensatory aforestation and cost of rehabilation and resettlement.
8			Total cost is the cost per unit of production.
	Category : Project Attributes;		Component : Investment Returns
1	Land Aquisition problems	0-1	Parameter measurement is subjective. Enter on a scale 0 to 1 where 0 corresponds to serious land aquisition problems while 1 corresponds to minimal problems. The land acquisition problems may arise due multiple land ownership, aquisition of land belonging to government departments such as railways and change in land use classification.
2	Labou Problems	0-1	Parameter nleasurement is subjective. Enter on a scale 0 to 1 where 0 corresponds to serious land aquisition problems while 1 corresponds to availability of reliable and efficient labour locally.
3	Govcrnment Incentives	0-1	Parameter measurement is subjective. Enter on a scale 0 to 1 where 0 corresponds to lack of motivation from the concerned state govrernment while 1 corresponds to willingness/ motivation on the part of the state government to promote establishing and operation of industries.

The major feature of ROSA is that environmental and project atrributes are expressed in commensurate units providing common basis for comparison. The steps involved in the development of commensurate units include transformation of parameter estimates into an Site Acceptibility Index (SAI) scale, assignment of importance units to the individual parameters (PIU), and the mul tiplication of scale values by importance values to obtain Site Acceptibility Score (SAS).

Thc overall score for the site alternative is then calculated as:

$$SAS_I = \sum_i \sum_j \sum_k (SAI)_{ijk} * (PIU)_{ijk}$$

in which

SAS1 = site acceptibility score for 1th site alternative

i = number of category

j = number of component corresponding to category i

k = number of parameter corresponding to component j

SAI = site acceptibility index corresponding to parameter measurement

PIU = parameter importance unit or weight assigncd to the parameter

Transformation of parameter estimates into an SAI scale is based on the range of anticipated values for a given parameter. For example, depth of ground water for economic abstraction will typically range from 5 to 100 metres. To help transforming the parameter estimates into an SAI scale, Site Acceptibility Curves (SACS) are developed with parameter values on abscissa (5 to 100 m depth of water table), and the SAI scale on the ordinate (0 to 1, with 0 representing very poor acceptibility and 1 representing excellent acceptibility). Assignment of importance units to each individual parameter is based on ranked pairwise comparison technique in which subjective judgment determines the relative importance or significance of this allocation. The technique is illustrated in Table 3.4. The methodology for development of SAC is summarised in Table 3.5.

Table 3.4. Ranked Painvise Comparison Technique

Assigning Parameter Importance Units (1'111) warrants a cross-section of persons to indicate their judgment on the relative importance of environmental impacts. Estimation of the relative importance of parameters, components and categories of environmental quality by assigning them PIC involves nine steps. These steps are outlined herein and also illustrated~witha numerical example.

1 Sclect a group of individuals for evaluation and explain to them in detail the PIU concept and the use of their rankings and PIUs.

2 Repeat steps 3-7 unit1 all PILJs have been distributed to the parameters.

3 Rank the parameters that are to be evaluated

4 Assign a value 1 to the first parameter on the list. Then compare the second parameter with the first to determine how much the second is worth compared to the first. Express this value as a decimal between 0 and 1.

5 Continue with these pair-wise comparisons until all parameters in the list have'been evaluated !compare 3 with 2.4 and 3, etc.)

6 Multiply out percentages and express over a common denominator.

7 Average these decimal values for all individuals. Multlply these averages by the PIUs to be distributed to these parameters.

8 Indicate Lo the group of individuals the results of their selections by using a test example. If they desire to change their selections, reevaluate all the weights. (Start with Step 2)

9 Repeat the experiment with same group or another group to increase the reliability of the results

Example :

1-2 Select three parameters (A,B,C) and repeat steps 3-9

3 Ranking of parameters - B,C,A

4 Assign weights B = 1

5 C = W2 of value of B

A = 1/2 of value of C

6 Multiply out percentages and express over common denominator

B = l

C = .5

A = .25

1.75

B = W1.75 = .57

C = .5/1.75 = .29

A = .25/1.75 = 1.4

7 Average the values of importance of (B=.67, C=.29, A=.14) over all other individuals in group. If these were the averages and 100 points were to be distributed between A, B, and C, then

B = .57 x 100 = 57 units

C = .29 x 100 = 29 units

A = .14 x 100 = 14 units

8-9 Continue with these steps to obtain reliable estimates of A, B, and C.

Table 3.5. Methodology for Development of Site Acceptability Curves

Delphi technique for plotting the site acceptability curves (SAC) is a commonly used procedure. SAC are plotted to bring out the relationship between a parameter measurement and the site acceptibility.

Seven steps are basically followed to plot the SAC

i Obain scientific information when available on the relationship between the parameter and the site acceptibility. Also, obtain experts in the field to develop the SACs.

ii Order the parameter scale so that the lowest value of the parameter measurement is zero and it increases in the positive direction.

iii Divide the acceptibility scale '(0-1) into equal intervals and express the relationship between an interval and the parameter measurement. Continue this procedure until a curve is completed.

iv Average the curves over all experts in the dxperiment to obtain a group curve (for parameters based solely on judgement, SACs should be determined by a representative population cross section).

v Indicate to the experts doing the SAC and expected results of using the curves in the ROSA. Decide if a modification is desired, if needed go to step (iii), if not continue.

vi Do step (i-v) until (iii) curves are developed for all parameters.

vii Repeat experiment with the same group or another group of persons to increase the reliability.

The SAC developed in the present research are based on literature, and interaction with experts. Typical SACs are presented in Figures 3.3 through 3.6. The numerical database for SACs is summarised in Table 3.6.

**The overall methodology for ROSA involves following steps:

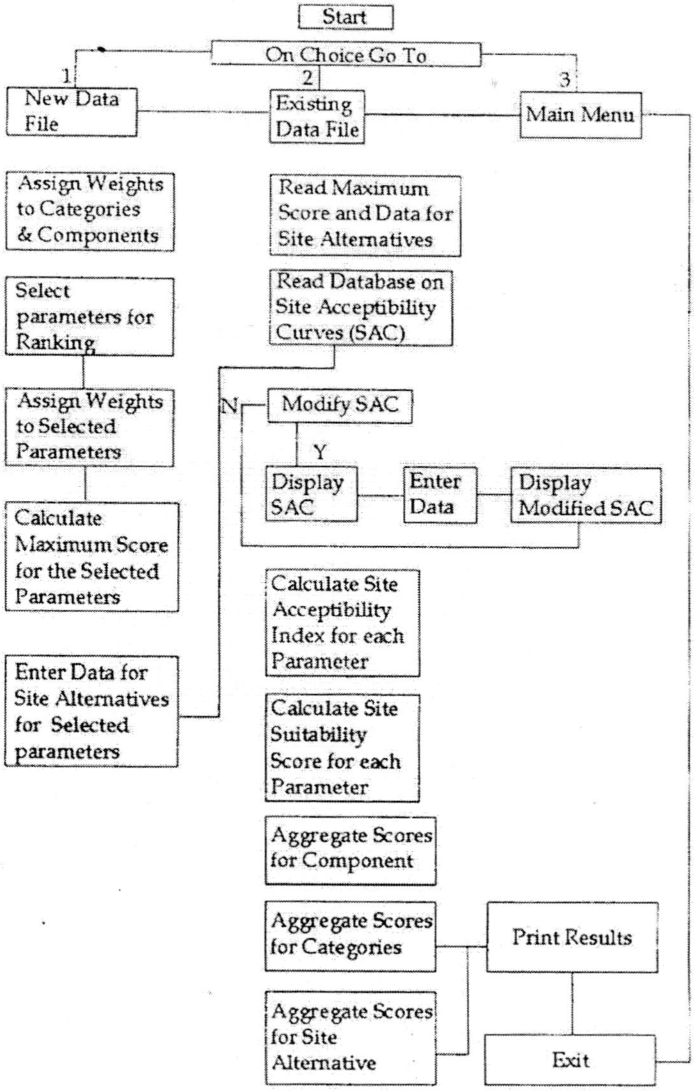

Fig. 3.3. System Flow-chart for ROSA : Ranking of Site Alternatives

Fig. 3.4. Questionnaire checklist developed for Impact Identification of Industrial Projects

1	Whether following type of land is proposed to he acquired for project'?		
	Agricultural	Yes	No
	Grazing	Yes	No
	Fallow	Yes	No
	Forest	Yes	No
	Mangroves	Yes	No
	Marshes	Yes	No
	Coral reafs	Yes	No
	Water sheds	Yes	No
	Industrial	Yes	No
	Land reserved for Settlement	Yes	No
	Green belt	Yes	No
	Park	Yes	No
	Recreational facilities	Yes	No
2	Does site belong to low lying area	Yes	No
3	Wheather following type of landuse is within 1 km radial distance?		
	Residential	Yes	No
	Commercial	Yes	No
	Industrial	Yes	No
	Agricultural	Yes	No
	Forest	Yes	No
4	Whether project would require water for		
	Process	Yes	No
	Cooling	Yes	No
	Domestic	Yes	No
5	What is the source of water supply		
	Surface water	Yes	No
	Ground water	Yes	No
6	Whether following units from part of raw water treatment facilities proposed at the project site		
	Sedimentation	Yes	No
	Chemical treatment	Yes	No
	Filtration	Yes	No
	Chlorination	Yes	No
	Demineralization	Yes	No
7	Whether following units frorn part of wastewater treatment facilities proposed at the project site		
	Oil & grease removal	Yes	No
	Equaliation	Yes	No
	Sedimentation	Yes	No
	Biological treatment	Yes	No
	Tertiary treatlnent	Yes	No

8. Whaf is propsed mode of watewater disposal

Water body	Yes	No
Land	Yes	No
Sewer	Yes	No

9. What is the source of power for the proposed project?

Captive power generation	Yes	No
Public supply	Yes	No
Thermal power plant	Yes	No
Hydel power plant	Yes	No
Nuclear power plant	Yes	No

10. What is fuel proposed to be used ?

Coal	Yes	No
Natural gas	Yes	No
LSHS	Yes	No
HSD	Yes	No
Naptha	Yes	No
Furnace oil	Yes	No
Wood	Yes	No
Bagasse	Yes	No
Rice husk	Yes	No
Agricultural wastes	Yes	No

11. Whether following unit form part of proposed air pollution control system at the project site

Bag filters	Yes	No
Electrostatic precipitators	Yes	No
Cyclone	Yes	No
Dry scrubbers	Yes	No
Wet scrubbers	Yes	No

12. Whether following source of noise would be present in proposed plant

Pneumatic power tools	Yes	No
Molding machine	Yes	No
Air blow down devices	Yes	No
Blowers	Yes	No
Air compressor	Yes	No
Metal forming	Yes	No
Cornbution	Yes	No
Turbo generators	Yes	No
Pumps	Yes	No
Trucks	Yes	No

13. Whether following type of solid wastes would be generated from proposed unit

Organic wastes	Yes	No
Inorganic wastes	Yes	No
Hazardous wastes	Yes	No

14. What is the proposed mode of disposal of solid waters

	Incineration	Yes	No
	Landfilling	Yes	No
	Municipal collection system	Yes	No
15.	What is the labour requirement of proposed industrial unit		
	Skilled	Yes	No
	Unskilled	Yes	No
	Other staff	No	Yes
16.	Whether housing facilities are proposed to be provideti by the industrial unit? -	Yes	No

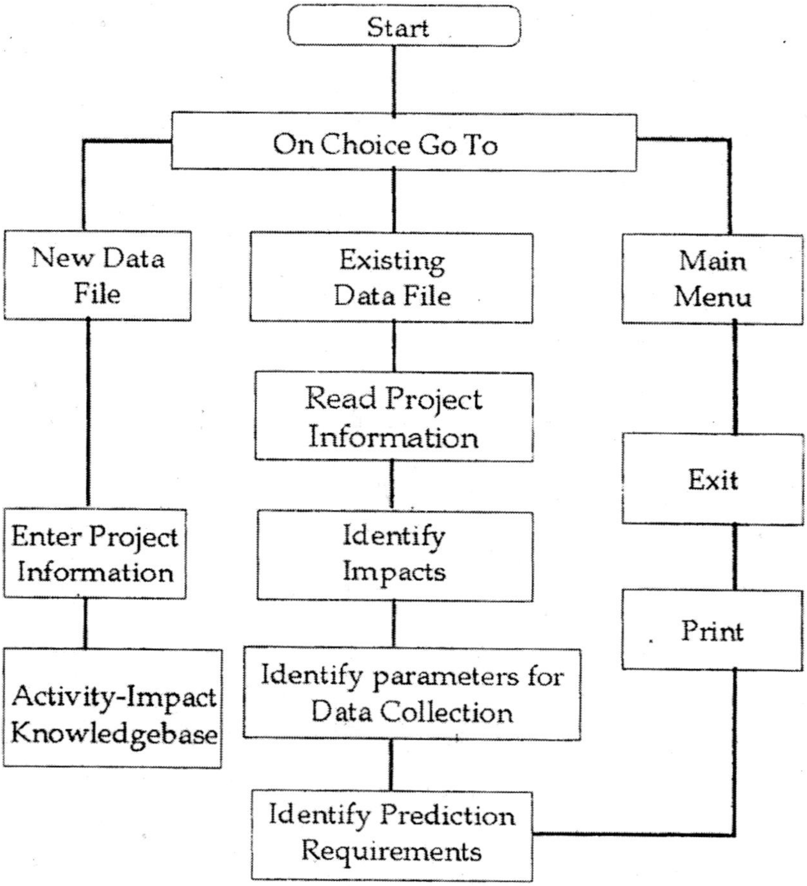

Fig. 3.5. System flow-chart for IMPACT Identification

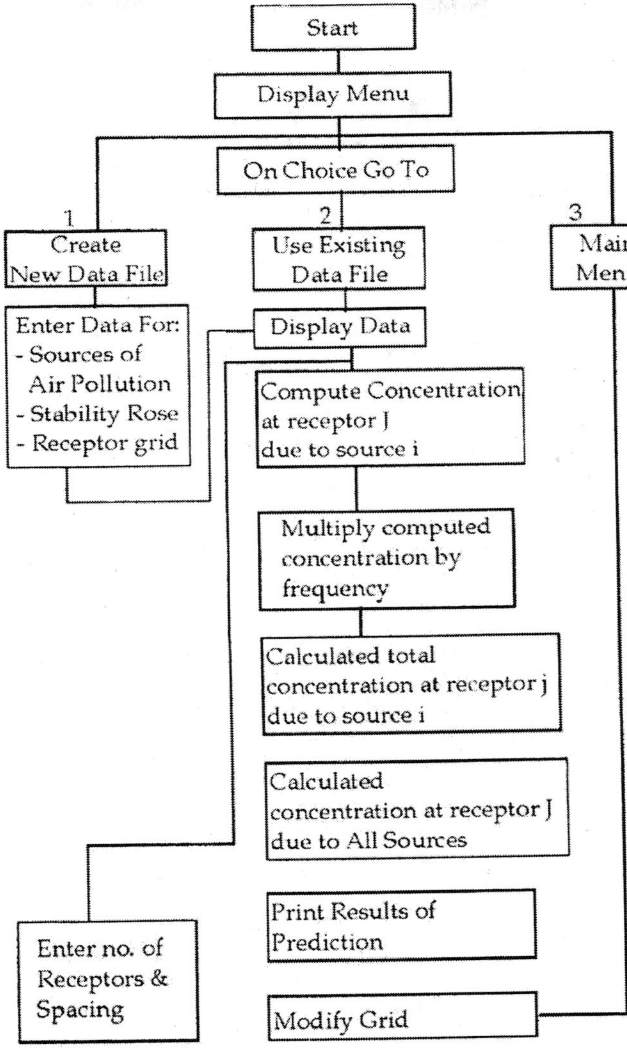

Fig. 3.6. System flow-chart for AQLONG : Long-term Prediction of Air Quality

Preparatory study

i. Identify a team of experts for site selection

ii Select parameters for evaluation of site alternatives for the project under consideration

Table 3.6. Data for development of site acceptibility curves

Sr. No.	Parameter	Measurement Unit	Site Acceptability Index										
			0.0	0.1	0.2	0.3	0.4	0.5	0.6	0.7	0.8	0.9	1.0
	Category : Supportive Capacity Component : Water												
1	Surface Water Availability	0-1	0.0	0.1	0.2	0.3	0.4	0.5	0.6	0.7	0.8	0.9	1.0
2	Ground water Depth	m	100.0	90.0	80.0	70.0	60.0	50.0	40.0	30.0	20.0	10.0	5.0
3	Ground water requirement	demand/yield	40.0	38.0	36.0	34.0	32.0	30.0	28.0	26.0	24.0	22.0	20.0
	Component : Power												
1	Power Requirement	demand/avail. %	75.0	70.0	65.0	60.0	55.0	50.0	45.0	40.0	35.0	30.0	25.0
2	Power Source	Poll. Pot. (0-1)	1.0	0.9	0.8	0.7	0.6	0.5	0.4	0.3	0.2	0.1	0.0
	Component : Infrastructure												
1	Road Transport Highway	dist. km.	75.0	50.0	45.0	40.0	35.0	30.0	25.0	20.0	15.0	10.0	5.0
2	Rail Transport - Station	dist. km.	75.0	50.0	45.0	40.0	35.0	30.0	25.0	20.0	15.0	10.0	5.0
3	Port/Harbour	dist. km.	200.0	150.0	100.0	75.0	60.0	50.0	45.0	40.0	35.0	30.0	25.0
4	Housing Facilities	0-1	0.0	0.1	0.2	0.3	0.4	0.5	0.6	0.7	0.8	0.9	1.0
5	Medical Facilities	0-1	0.0	0.1	0.2	0.3	0.4	0.5	0.6	0.7	0.8	0.9	1.0
6	Skilled Labour	0-1	0.0	0.1	0.2	0.3	0.4	0.5	0.6	0.7	0.8	0.9	1.0
	Component : Land												
1	Forest Land for Project	%	10.0	9.0	8.0	7.0	6.0	5.0	4.0	3.0	2.0	1.0	0.0
2	Agri. Land for Project	%	20.0	18.0	16.0	14.0	12.0	10.0	8.0	6.0	4.0	2.0	0.0
3.	Waste Land for Project	%	0.0	10.0	20.0	30.0	40.0	50.0	60.0	70.0	80.0	90.0	100.0

Table 3.6 contd...

	Category : Assimilative Capacity												
	Component : Ambient												
1	Susp. Part Matter	ug/cum	500.0	300.0	225.0	150.0	125.0	100.0	75.0	50.0	25.0	10.0	0.0
2	Sulphur Dioxide	ug/cum	1310.0	655.0	420.0	315.0	236.0	197.0	157.0	131.0	105.0	65.0	0.0
3	Nitrogen Dioxide	ug/cum	1230.0	209.0	80.0	61.0	55.0	50.0	37.0	34.0	30.0	18.0	0.0
4	Carbon Monoxide	ppm	45.0	42.0	30.0	25.0	22.0	19.0	17.0	15.0	14.0	5.0	0.0
5	Hydrocarbons	ppm	0.5	0.4	0.2	0.2	0.2	0.2	0.2	0.1	0.1	0.1	0.0
6	Photochemical Oxidants	ug/com	1.0	0.3	0.1	0.0	0.0	0.0	0.0	0.0	0.0	0.0	0.0
	Component : Ambient Water Quality												
1	Dissolved Oxygen	mg/lit	0.0	2.0	3.0	3.5	4.2	5.0	5.5	6.0	6.5	7.5	9.0
2	Biochem. Oxygen Demand	mg/lit	20.0	15.0	10.0	8.0	7.0	6.0	5.0	4.0	3.0	2.0	0.0
3	PH	deviation from7	5.0	3.0	2.5	2.2	1.8	1.4	1.2	1.1	1.0	0.5	0.0
4	Temperature	deviation Oc	10.0	5.0	4.5	4.0	3.5	3.0	2.5	1.8	1.0	0.5	0.0
5	Faecal Coliform	MPN/100 ml	100000	10000	1000	200.0	100.0	30.0	20.0	10.0	5.0	3.0	0.0
6	Total Phosphates	mg/lit	10.0	5.0	2.5	1.5	1.0	0.8	0.6	0.4	0.2	0.1	0.0
7	Total Nitrates	mg/l	100.0	55.0	35.0	22.0	15.0	10.0	7.5	5.0	2.5	1.0	0.0
8	Turbidity	JTU	200.0	150.0	90.0	65.0	50.0	35.0	20.0	15.0	9.0	5.0	0.0
9	Total Solids	mg/l	1000.0	700.0	600.0	500.0	450.0	375.0	300.0	225.0	150.0	150.0	150.0
	Component : Ambient Noise Levels												
1	Noise Level - Sprace	max.dB	100.0	95.0	90.0	85.0	80.0	75.0	70.0	65.0	60.0	55.0	50.0
2	Noise level - frequent	max.dB	100.0	92.0	85.0	75.0	70.0	65.0	62.0	59.0	57.0	55.0	50.0
3	Nosie Level - continuous	max.dB	100.0	85.0	75.0	70.0	65.0	61.0	57.0	55.0	53.0	51.0	50.0
	Component : Soil Quality												
1	Exchangeable Sodium %	ESP	15.0	14.0	13.0	12.0	11.0	10.0	8.0	6.0	4.0	2.0	0.0

Table 3.6 contd...

2	Water Extractable Metal	mg/l	2.0	1.8	1.6	1.4	1.2	1.0	0.8	0.6	0.4	0.2	0.0
	Component : Air Assimilative Capacity												
1	Topography	0-1	0.0	0.1	0.2	0.3	0.4	0.5	0.6	0.7	0.8	0.9	1.0
2	Wind speed	m/s	1.0	1.5	2.0	2.5	3.0	3.5	4.0	4.5	5.0	5.5	6.0
3	Wind Dir. towards Receptor	%	70.0	'0.0	50.0	40.0	30.0	25.0	20.0	15.0	10.0	5.0	0.0
4	Inversions	nos/yr	10.0	9.0	8.0	7.0	6.0	5.0	4.0	3.0	2.0	1.0	0.0
5	Mixing Depth	m.	1000	1500	2000	2250	2500	2750	3000	3250	3500	4000	4500
6	Ventilation Coeff.	m2/s	6000	7000	8000	9000	10000	11000	12000	13000	14000	15000	16000
	Component : Water Assimilative Capacity												
1	Dilution Potential	ratio	1.0	2.0	3.0	4.0	5.0	6.0	7.0	8.0	9.0	10.0	20.0
2	Flow Velocity	m/s	0.1	0.2	0.3	0.4	0.5	0.6	0.7	0.8	0.9	1.0	1.5
	Component : Soil Assimilative Capacity												
1	Soil Permiability	% clay	0.0	10.0	20.0	30.0	40.0	50.0	60.0	70.0	80.0	90.0	100.0
	Component : Ecology												
1	Distance to Historic Site	km.	25.0	30.0	35.0	40.0	45.0	50.0	55.0	60.0	65.0	70.0	75.0
2	Distance to Archio. site	km.	25.0	30.0	35.0	40.0	45.0	50.0	55.0	60.0	65.0	70.0	75.0
3	Distance to Bio. Reserves	km.	25.0	30.0	35.0	40.0	45.0	50.0	55.0	60.0	65.0	70.0	75.0
4	Distance to National Park	km.	25.0	30.0	35.0	40.0	45.0	50.0	55.0	60.0	65.0	70.0	75.0
5	Distance to Sanctuaries	km.	25.0	30.0	35.0	40.0	45.0	50.0	55.0	60.0	65.0	70.0	75.0
	Category : Human Interest Component : Aesthetics												
1	Landscape	0-1.0	1.0	0.9	0.8	0.7	0.6	0.5	0.4	0.3	0.2	0.1	0.0
2	Vegetation Cover	% Pr. Area	100.0	90.0	80.0	70.0	60.0	50.0	40.0	30.0	20.0	10.0	0.0

Table 3.6 contd...

	Component : Socio Economics												
1	Work Participation	ratio	1.0	0.9	0.8	0.7	0.6	0.5	0.4	0.3	0.2	0.1	0.0
2	Community Health	0-1.0	0.0	0.1	0.2	0.3	0.4	0.5	0.6	0.7	0.8	0.9	1.0
3	Population Density	pop./sq. km	1000	800.0	600.0	400.0	300.0	250.0	200.0	175.0	150.0	125.0	100.0
4	Per Capita Income	Rs/yr	7000	6500	6000	5500	5000	4500	4000	3500	3000	2500	2000
	Category : project Attributes Component : Production Cost												
1	Transport	0-1	0.0	0.1	0.2	0.3	0.4	0.5	0.6	0.7	0.8	0.9	1.0
2	Water	0-1	0.0	0.1	0.2	0.3	0.4	0.5	0.6	0.7	0.8	0.9	1.0
3	Power	0-1	0.0	0.1	0.2	0.3	0.4	0.5	0.6	0.7	0.8	0.9	1.0
4	Labour	0-1	0.0	0.1	0.2	0.3	0.4	0.5	0.6	0.7	0.8	0.9	1.0
5	Land	0-1	0.0	0.1	0.2	0.3	0.4	0.5	0.6	0.7	0.8	0.9	1.0
6	Construction	0-1	0.0	0.1	0.2	0.3	0.4	0.5	0.6	0.7	0.8	0.9	1.0
7	Env. Management	0-1	0.0	0.1	0.2	0.3	0.4	0.5	0.6	0.7	0.8	0.9	1.0
8	Total Cost	0-1	0.0	0.1	0.2	0.3	0.4	0.5	0.6	0.7	0.8	0.9	1.0
	Component : Investment Returns												
1	Land Aquisition Problems	0-1	1.0	0.9	0.8	0.7	0.6	0.5	0.4	0.3	0.2	0.1	0.0
2	Labour Problems	0-1	1.0	0.9	0.8	0.7	0.6	0.5	0.4	0.3	0.2	0.1	0.0
3	Govt. Incentives	0-1	0.0	0.1	0.2	0.3	0.4	0.5	0.6	0.7	0.8	0.9	1.0

iii. Decide the site acceptability indices (SAI) for the selected parameters corresponding to range of f'casiblc mcasure-ments and develop Site Acceptability Curves (SACS) using Delphi Technique. The Delphi Technique is described in Table 3.7

iv. Apportion a total scorc (weightage) of 1000 betwccn the assessment parameters (Parameter Importance Units (PIU)) based on their importance through ranked pairwise comparison technique

Table 3.7. Delphi Technique

Delphi is a procedure to obtain the most reliable consensus of opinion of a group of experts by a series of intensive qucstionnaires intersprcsed with controlled opinion feedback.

In the key element Delphi process are :

i structuring of information flow

ii feedback to the participants, and,

iii Anonymity for the participants.

The method involved the following ten steps:

1 Formation of a team to undertake and monitor a Delphi on a given subjekt.

2 Selection of one or more panels to participate in the exercise (customarily, the panelists are experts in the area to he investigated)

3 Development of the first round Delphi cluestionnaire

4 Testing of the questionnaire for proper working (c.g. ambiguities, vagueness).

5 Transmission of the first questionnaires to the panelists.

6 Analysis of the first round responses

7 Prepration of the second round questionnaries (and possible testing)

8 Transmission of the second round questionnair to the panelists.

9 Analysis of second round responses (steps 7 to 9 are reiterated as long as desired or as necessary to achieve stability in the results)

10 Preparation of a report by the analysis stiesa m to present the conclusion of the exercise.

At-site study

v. Collect baseline data for the selected parameters for all site alternatives based on documented information and preliminary surveys, if necessary. No elaborate field data collection is expected at this stage. Also, the impact prediction for site selection studies is based on expert judgment

Site acceptability computations :

vi. Repeat steps (vii) through (x) for all site alternatives (j)

vii. Repeat steps (viii) and (ix) for all parameters (i)

viii. Using data collected in step (v) for the parameter (i) and site suitability curve developed in step (iii) SACi, estimate Site Acceptibility Index for the parameter as SAIi

ix. Calculate site acceptability score for the parameter (SASi) by multiplying SAIi with parameter importance units PIUi assigned in step (iv) as

SASi = SAIi * PIUi

x. Add the SAS for individual site alternatives (j) as

SASj = $\sum$ SASi

xi. Rank the alternatives based on descending order of the total score SASj.

The site selection module in EIA-AID incorporates methodology developed for ROSA in the present research.

**Salient features of the ranking of site alternative module are :

- Assign weightages to categories and components through ranked pairwise comparison technique in an interactive mode
- Flexibility of modifying weights so assigned
- On-screen selection of parameters and assigning weights
- Default weights assigned for naive user
- Site acceptability curves stored for all parameters
- User can modify SAC as per his perception of the project and environment
- Option to modify scaling function database using text-editor
- MeasuremeAlt units defined on screen for easy data entry
- On screen confirmation for correct data entry in interactive mode
- Data for three sites can be entered at a time

Systems flowchart for ranking of site alternatives is shown in Fig. 3.3.

Fig. 3.7. Conlbinations of' variables to be considered for DO analysis in streams nnd rivers

Combination - I

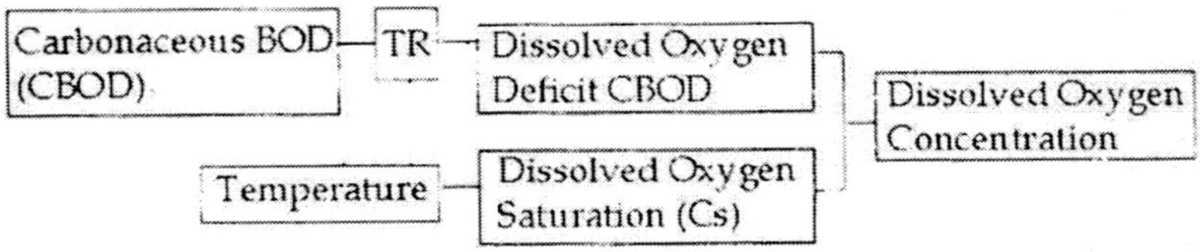

Combination - II

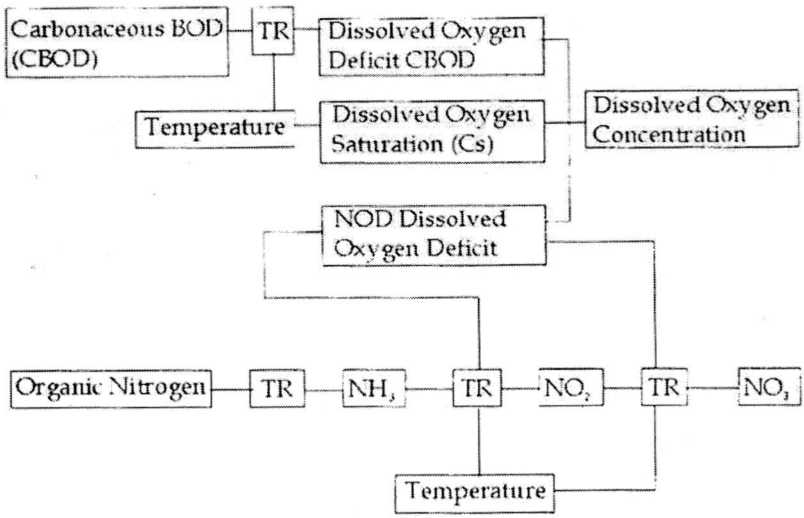

Combination - III

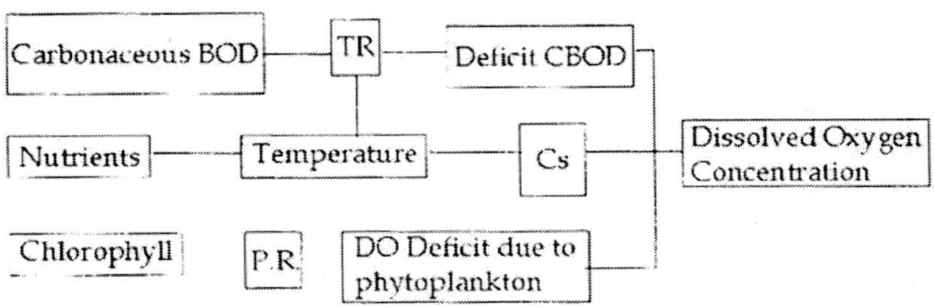

Fig. 3.7. Contd...

Combination - IV

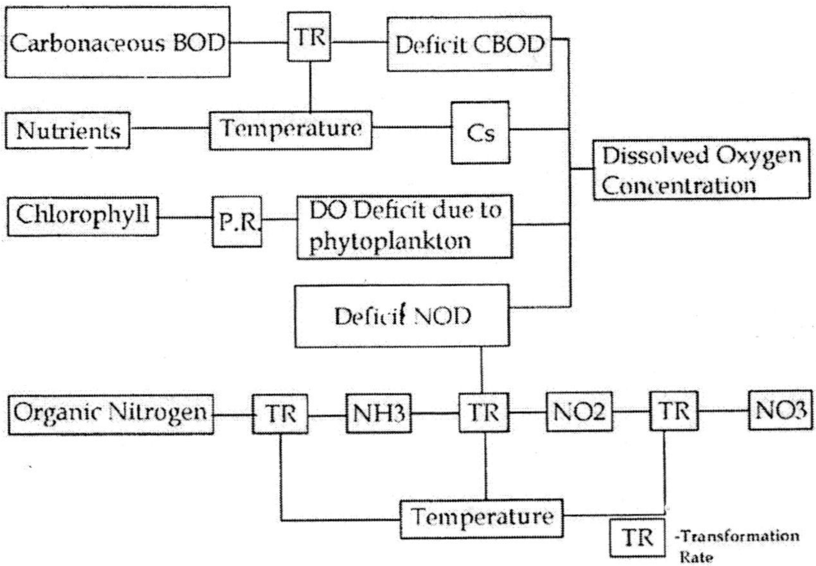

Fig. 3.7.

3.4. IMPACT IDENTIFICATION

Environmental impact is any alteration of environmental conditions or creation of a new set of environmental conditions, adverse of beneficial, caused or induced by the project under consideration. Impact on environment depends on the nature, scale, and location of the proposed activity.

Environmental impacts include effect on natural resource base; quality of air, water, noise, biological and socio-economic components of environment; effect on public health as also cost of environmental management.

The impacts can be classified as primary which can be attributed directly to the project, and secondary which are indirect or induced changes and typically include the changed patterns of socio-economic activities likely to he stimulated or induced by the proposed activity.

The most logical way of identifying impacts is the preparation of cause-condition-effect networks for various activities which form components of the proposed project. Impact identification results in selection of parameters for baseline data collection based on considerations of regulatory requirements, resources available, magnitude of the project, and prevailing environmental quality at the site. Eventhough the impact networks for a type of project may be the same, the interpretation of parameters is site specific, and requires substantial professional experience.

An extensive database on environmental impacts trigerred off by the industrial prqject related activities has been developed in the present research, and is presented in Tabln 3.8. In order to assist the EIA team in planning field data collection activities and identification of appropriate impact assessment (quantification) techniques, a knowledge base has been built-up relating project activities, data requirements for impact assessment, and impact measurement techniques (Table 3.9) in the present research based on the interactions with the subject experts at the National Environmental Engineering Research Institute (NEERI).

Table 3.8. Impact identification Database for EIA-AID

Actlvity	Primary impacl/ event	Seconary impact/ event	Tertiary impact/ event
Plant construction related activities (site cleaning to labour camps)	Change in air, water and land quality	Damage to huillan health & vegetation	Change in quality of life
Acquisition of agric- ultural land for project	Change in landuse pattern	Loss of agricultural production	Impact on economic output of region
Acquisition of graz- ing land for project	Change in landuse pattern	Reduction in fodder availability	Increase in cost of fodder
		Loss of forest products	
		Reduction in precipitation	
Acquisition of man- grove land for project	Disturbance to fragile ecosystem	Loss of flora & fauna	Extintion of rare & endangered species
Acquisition of land in watershed	Change in runoff pattern	Reduction in water impoundment capacity	Decrease in water availability

	Increase in soil erosion		
Acquisition of notified industrial land for project	Increase in polltrtion load per unit area	Reduction in pollution assimilation potential	Chance in quality of life parameters in nearby areas
Acquistion of land reserved for welfare activities	Impact on landuse plants	Conjection of anthropogenic activities	Change in quality of life
Acquisition of low lying land for project	Levelling of land by filling	Quariying for fill material	Transport of fill material
Wateruse in process	cooling/domestic purposes	Generation of wastewater	
Abstraction of surface water	Change in water availability in water bod; DIS of withdrawal point	Reduction in dillution aotential	Loss of aquatic flora & fauna Change in hydraulics Increase in cost of public water supply schemes Reduction in agricu ltural production
Abstraction of ground	Lowering of water table	Increase In cost of abstraction Changes in hydrogeology Change in on water quality	Decrease in supportive capacity for water
Raw water treatment to meet quality requirements (sedimentation/chemical treatment/ filtration)	Generation of solid wastes (sludges)	Impact on water/ soil quality due to sludge disposal	Impact on ground water quality
Raw water treatment to meet quality requirements(dimeneralizationl reverse osmosis)	Generation of back- wash/reject water	Load on wastewater treatment plant design	Increase in cost of production
Wastewater treatmerit to meet effluent standards (preli minary/ primary/ secondary treatment)	Generation of solid wastes (sludges)	Inlqact on water/ soil quality due to sludge disposal	Impact on ground water quality

Wastewater treatment to meet effluent standards (ahsorbtion/ dimeneralizatiod/ reverse osmosis)	Generation of back- washlreject water	Load on wastewater treatment plant design	Increase in cost of production
Discharge of wastewater in water body	Change in water quality of receiving waters	Loss of aquatic flora & fauna	Cost of water treatment a t down stream
		Impact on designated use on water body	
Discharge of waste-water on land	Change in soil and ground water quality	Chagne in soil fertility	Impact on agricultural output
		Impact on health of farm workers & water consumers	Impact on cost of water treatmen
Discharge to municipal sewars	Increase in pollution load to municipal treatment plant	Impact on plant design/ performance	Impact on cost of treatment
Abstraction of power from public power supply grids	Change in power consumption pattern in region	Impact on economic output of region	Impact on operation of power plantdgrid
Co-generation of power	Change in environmental quality		
Use of coal as fuel	Emission of SPhl, SOX, NOX, HC	Change in air quality	Impact on human health, vegetation and material
		Deposition of particulates	Impact on soil/ water quality
Cse of natural gas/ LSHS/HSD/Furnace oil as fuel	Emission of SOX: NOX, HC, CO	Change in air quality	Iinpact on humari health, vegetation and material
Air pollution control units to meet standards (fabric filters/ESP/ scrubbers cyclones)	Generation of solid/ liquid wasks	Impact on water/ soil quality due to waste disposal	Impact on ground water quality
Emission of noise	Increase in ambient noise levels	Damage to human health and fauna	Impact on economic output
Incineration of solid / hazardous wastes	Emission of SPM, SOX, NOX, HC, CO, VOC	Change in air quality	Impact on human health, vegetation and material
		Deposition of particulatcs	Impact on soil/ water quality

Disposal of solid/ hazardous waste on land	Change in soil and ground water quality	Change in soil fertility	Impact on human health
			Impact on agricult-ural output
			Impact on agricultural output
Employment of manpower	Change in socio economic environment	Impact on purchasing power	Impact on economic output
	Impact on infrastructure facilities	Economic input for Infrastructure facilities	
Development of housing facilities	Impact on socio economic environment	Impact on infras-tructure facilities	Impact on economic output

Table 3.9 Impact measurement techniques Database in EIA-AID

Activity	Data Requirements for EIA	Impact Measurement
Plant construction related activities (site cleaning to labour camps)	Sources of constr. material transport routes Labour requirement sits accoinodation	Impact of transport construction waste disposal
		Socioeconomic impads of labour population
Acqisition of agricultural land for project	Baseline data on land productivity	Economic loss due to lost crop yield
	Baseline data on supportive capacity for cattle	Reduction in supportive capacity for cattle
Acqisition of forest land for project	Baseline data on terrestrial ecology	Impact on species diversity rare & endangered species
	Forest products forest clearance procedure	Loss of forest products
Acqisition of mangrove land for project	Baseline data on mangrove ecosystem	Impact on species diversity rare & endangered species
	Baseline data on mangrove products	Impact on mangrove production
Acqisition of land in watershed	Baseline data on rainfall runoff pattern	Impact on runoff pattern and soil erosion
	Baseline data on soil erosion	
Acqisition of notified industrial land for project	Inventory of existing sources of pollution	Aggregated impact on quality of life

Acqisition of land reserved for welfare activities	Inventory of existing areas for welfare activities	Impact on quality of life
Acqisition of low lying and for project	Topography of quarrying site	Impact of quarrying and transport and compaction of fill material
	Distance to quarrying site	
	Mode of transport	
	Land filling procedure	
Water use in process cooling1 domestic purpose	Water balance in the proposed unit	Feasibility of waste minimi zation, recyclelreuse, bio-methanation
Abstraction of surface water	Competing water demand for various uses.	Water availability for various uses
	Water quality inventory of wastewater discharge	Impact on surface water quantity & quality
Abstraction of ground water	Baseline depth of water table regenerative	
	Capacity of aquifer competing demands	
Raw water treatment to meet quality requirements (sedimentation/chemical treatment/ filtration)	Raw wter quality vls industry requirement	Change in waterlsoil quality due to sludge disposal
	Details of treatment plant & proposed sludge handling practice	
Raw water treatment to meet quality requirements (dimeneralizatiod reverse osmosis)	Quantity of backwash rejects	Efficacy of treatment method
	Proposed treatment of backwash/rejects	
Wastewater treatment to meet effluent standards (preliminary/primary/sec ondaly treatment)	Wastewater quality vls standards	Change in water/soil quality due to sludge disposal
	Details of treatment plant & proposed	
	Sludge handling practice	
Wastewater treatment to meet standards (absorption demineralization/ reverse osmosis)	Quantity of backwash/ rejects	Efficacy of treatment method
	Proposed treatment of backwash/rejects	Change in water quality
Discharge of wastewater in water body	Baseline water quality and designated use	
	Inventory of wastewater discharge (existing & new)	

	Hydraulics of water body materiology	
Discharge of wastewater on land	Baseline data on soil and ground water quality	Change in soil and ground water quality
	Treated/untreated waste-water loads from proposed unit	
Discharge to municipal sewers	Existing wastewater treatment facilities	Excepted problems in mun icipal treatment facilities
	Treated/untreated waste-water loads from proposed unit	
Abstraction of power from public power supply grids	Competing demands v/s power availability	Impact on peak demand & power reliability
	Future plans for power generation	
Co-generation of power	Baseline environmental quality	Analyse as a separate proposal
Use of coal as fuel	Baseline air quality within 10 kin distance	Increase in concentration of air pollutants in the impact zone
	Meterological data emission from proposeu unit	
	Inventory of pollution sources in region	
Use of natural gas/ LSHS HSDI Furnace oil as fuel	Baseline air quality within 10 km distance	Increase in concentration of air pollutants in the impact zone
	Meterological data emission from proposed unit	
	Inventory of pollution sources in region	
Air pollution control units to meet standards (fabric filters/ESP/scrubbers/cyclones)	Emission characteristics v/s standards	Change in water/soil quality due to waste disposal
	Quantity of secondary wastes generated	Efficacy of pollution control units
	Details of pollution control units & proposed secondary waste handling practice	

Emission of noise	Baseline noise levels in impact zone	Increase in noise levels in impact zone
Incineration of solid hazardous wastes	Baseline air quality within 10 km distance Meterological data emissions from proposed data	Increase in concentration of air pollutants in the impact zone.
	Inventory of pollution sources in region	
Disposal of solid / hazardous waste on land	Baseline data on soil and ground water quality Treated/untreated waste loads from proposed unit design of landfill	Change in soil and ground water quality
Employment of manpower	Baseline data on socio economic status	Change in quality of life
Development of housing facilities	Baseline data on infrastructure facilities Demand for infrastructure facilities	Change in quality of life

The emission standards to be complied by the industries in India (Table 3.10) as also guidelines from the Bureau of Indian Standards for environmental management (Table 3.11) as a part of the present research. This information forms the knowledge base in the module.

The impact identification module in EIA-AID interacts with the user to retrieve information about the project then uses the knowledge base stored in the programme to identify primary, secondary and tertiary impacts associated with the proposed activity. It then delineates parameters for baseline data collection and prediction. It also indicates agencies which routinely collect the type of data required to assess the impacts, and delineates impact assessment parameters. If desired, the module provides the user with a list of Emission Standards and Indian Standards for environmental monitoring. Questionnaire check list that forms input in impact identification module is presented in Fig. 3.4. Systems flowchart for impact identification is presented in Fig. 3.5.

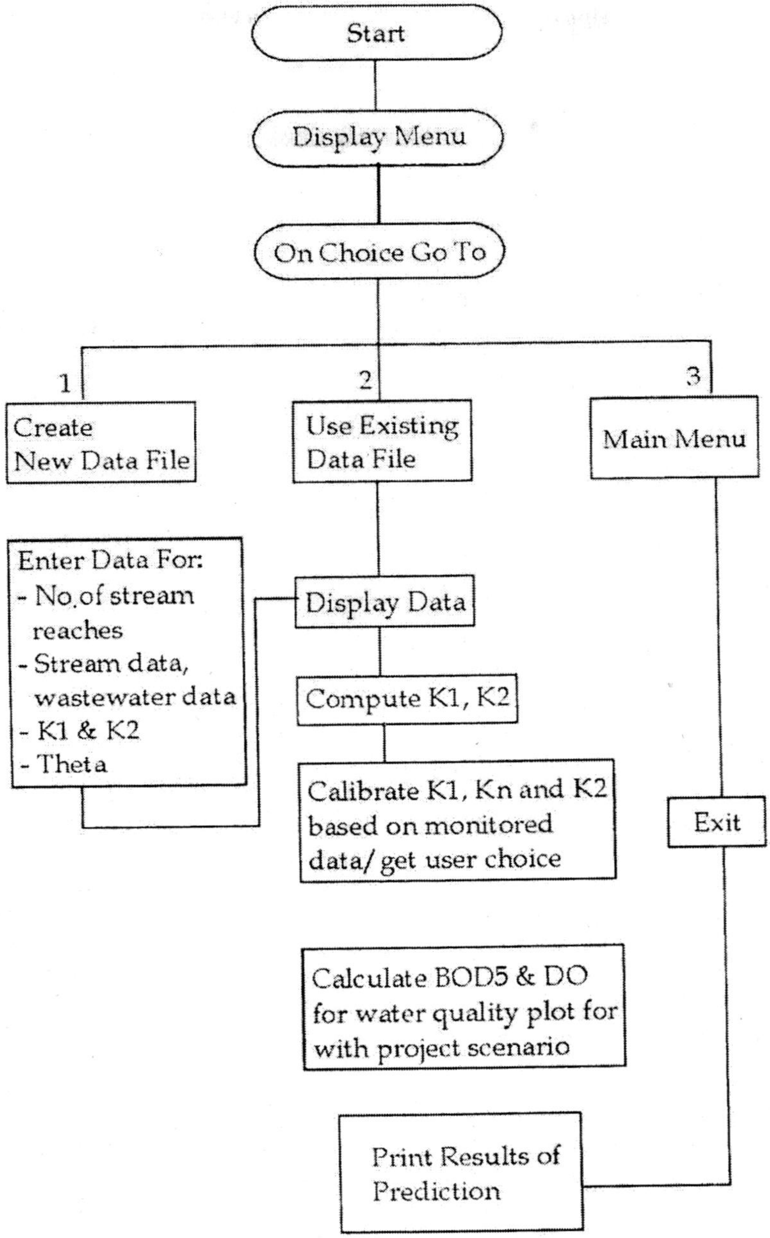

Fig. 3.8. System Flow-chart for WQMODEL - Prediction of Water Quality

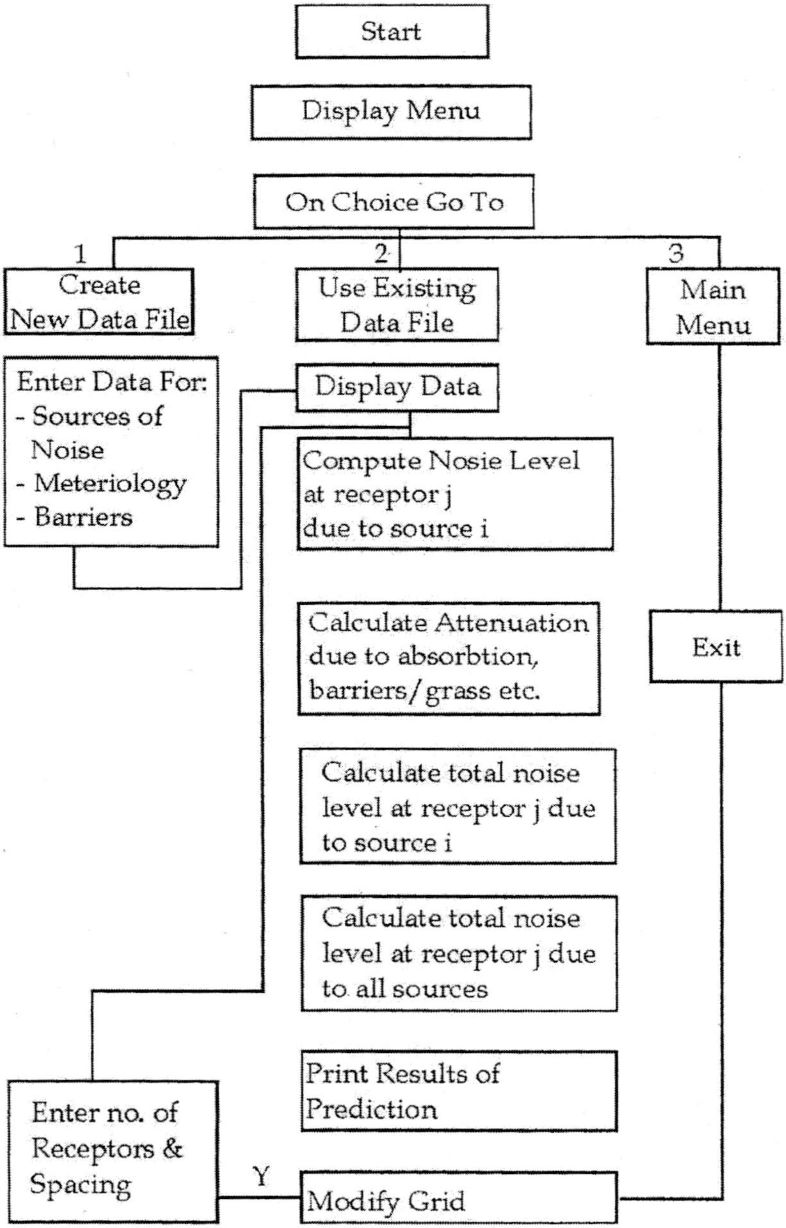

Fig. 3.9. System Flow-chart for NQMODEL - Noise Level Prediction

Table 3.10. Environmental Quality Standards in India

Standard	Coverage	Publishe
Gazette of India, May, 1993 (As per Environment Protection Act, 1986)	Emission & Standards – General	Ministry of Environment & Forest (MEF)
IS-2296	Surface water quality subjected to pollution	MEF
Gazette of India, Sept. 1988 (As per Environment Protection Act, 1986)	Effluent standards – General	MEF
Ambient Air Quality Standard - Notification	Ambient Air Quality – Various Zones	Central Pollution Control Board (CPCB
Gazette of India, Dec. 1989 (As per Environment Protection Act, 1986)	Ambient air quality – standards for noise	MEF
Permissible level of certain chemical substances in work environment	TLV in work zone	Factories Act (Amendment of 1987) (also ACGIH norms) Central Labour Institute, Bombay
Permissible level of noise in work zone	Noise pressure level and exposure time	Central Labour Institute, Bombay (also OSHA norms)
Minimal National Standards (MINAS)	Limits of Effluents	CPCB
Emission Regulations	Emission limits and stock heights	CPCB

Table 3.11. Bureau of Indian Standards Codes for Environmental Management

IS No.	Year	Part	Title
			AIR QUALITY MANAGEMENT
5182			Methods for measurement of air pollution
5182	1969	I	Dustfall
5182	1969	II	Sulphur dioxide
5182	1970	III	Radiactivity (Particulate) in air
5182	1973	IV	Suspended matter
5182	1975	V	Sampling of gaseous pollutants
5182	1975	VI	Nitrogen oxides
5182	1973	VII	Hydrogen sulphide
5182	1976	VIII	Sulphation rate
5182	1974	IX	Oxidants
5182	1979	X	Carbon monoxide

5182	1974	XII	Polynuclear aromatic hydrocarbons In air particulate matter
5182	1979	XIV	Guidelines for planning the sampling of atmosphere
5182	1974	XV	Mass concentration of particulate matter in the atmosphere
5182	1980	XVI	Recommended practice for collection, filtration and determination of mass, number and optical sizing of atmospheric particulates
5182	1979	XVII	Guidelines for planning the sampling of atmosphere
5182	1974	XVIII	C1 to C5 hydrocarbons in air gas chromatography
1635	1977		Limits for gaseous emissions from sulphuric acid and phosphatic fertilizer industries
1636	1977		Limits for gaseous emissions from petroleum refineries
8829	1978		Guidelines for micrometerological techniques in air pollution studies
9005	1978		Limits for gaseous emissions form nitric acid and nitrogenous fertilizer industries
			NOISE MONITORING
3028	1965		Methods of measurements of noise emitted by inotor vehicles
3931	1966		Sound level meter for the measurement of noise emitted by motor vehicles
3932	1966		Sound level meters for general purpose use
3486	1966		Code of practice for noise reduction in industrial building
4242	1967		Methods of ineasureinent of acoustical noise emitted by ballasts for gaseous discharge lamps
4954	1968		Recommendations for noise abatement in town planning
4758	1968		Methods of measurenlent of noise emitted by machines
6098	1971		Methods of measurement of the air borne noise emitted by rotating electrical machinery
			SOIL CONSERVATION
6518	1972		Code of practice for control of sediment in reservoirs
			SOLID WASTE MANAGEMENT
9235	1979		Methods for physical analysis and determination of moisture in solid wastes (excluding industrial wastes)
9234	1979		Methods for preparation of solid waste sample for chemical and microbiological analysis
			WASTE WATER MANAGEMENT
2488	1966	I	Methods of salnpling and test for industrial effluehts
2488	1978	II	Methods of salnpling and test for industrial effluents
2488	1968	III	Methods of sampling and test for industrial effluents
2488	1974	IV	Methods of salnpling and test for industrial effluents
2488	1976	V	Methods of salnpling and test for industrial effluents
1172	1971	-	Code of basic requirement for water supply, drainage and sanitation (first revision)

3.5 PREDICTION OF IMPACTS

3.5.1 Prediction of Impacts on Air Quality

3.5.1.1 Problem Identification

Air pollution, as also concerns on air quality management have assumed significant dimensions in the last two decades in almost all countries. Industry is one of the major sources of air pollution (Tables 3.12 and 3.13) (EPA, 1976; Rau & Wooten, 1980). Use of fossil fuels for generation of steam and electrcity for indiustrial processes results in air pollution. Depending upon the fuel and combustion technology used, the emission factors as also spatial scale of impacts vary significantly as indicated in Tables 3.14 (Flavin and Lenssen, 1994) and 3.15. Residence time and concentration in abmbient air of select constituents is presented in Table 3.16 (Strauss, 1980).

Table 3.12. Anthropogenic Sources of Air Pollution

1.	External combustion sources	Bituminous coal combustion Anthracite coal combustion Fuel oil combustion Natural gas combustion Liquified petroleum gas consumption Wood-waste combustion in boilers Lignite combustion
2.	Solid waste disposal	Refuse incineration Automobile body incineration Conical burners Open burning Sewage sludge incineration
3.	Internal combustion engine sources	Highway vehicles Light-duty, gasoline-powered vehicles (auotomobiles) Light-duty, diesel-powered vehicles Light-duty, gasoline-powered trucks and heavyduty, gasoline-powered vehicles Heavy-duty, diesel-powered vehicles Gaseous-fueled vehicles Motorcycles Off-highway, mobile sources Aircraft Locomotives Inboard-powered vessels Outboard-powered vessesl Small, general utility engines Agricultural equipment Heavy-duty construction equipment Snowmobiles

		Off-highway stationary sources Stationary gas turbines for electric utility power plants Heavy-duty, natural gas-fired pipeline compressor engines Gasoline and diesel industrial engines
4	Evaporation loss sources	Dry cleaning Surface coating Pctroleum storage Gasoline marketing
5	Chemical process industry	Adipic acid Ammonia Carbon black Charcoal Chlor-alkali Explosives Hydrochloric acid Hydrofluoric acid Sitric acid Paint and varnish Phosphoric acid Phthali anhydride Plastics Printing ink Soap and detergents Sodium carbonate Sulfuric acid Elemental sulfur-burning plants Spent-acid and hydrogen sulfide burning plants Sulfide ores and smelter gas plants Sulfur Synthetic fibers Synthetic rubber Terephthalic acid
6	Food and agricultural industry	Alfalfa dehydrating Coffee roasting Cotton ginning Feed and grain mills and elevation Fermentation Fish processing Meat smokehouses Nitrate fertilizers Orchard heaters Phosphate fertilizers Starch manufacturing Sugar cane processing
7	Metallurgical Industry	Primary alurninium production Metallurgical coke manufacturing Copper smelters Ferroalloy production Iron and steel mills Lead smelting

		Zinc smelting Scondary aluminium operations Hrass and bronze ingots Gray lron foundry Secondary lead smelting Secondary magnesium smelting Steel foundaries Secondary zinc processing
8	Mineral products industry	Asphaltic concrete plants Asphalt roofing Bricks and related clay products Calcium carbibe manufacturing Castable refractories Portland cement manufacturing Ceramic clay manufacturing Clay and fly-ash sintering Coal cleaning Concrete batching Fiberglass manufacturing emissions & controls Frit manufacturing Glass manufacturing Gypsum manufacturing Lime manufacturing Mineral wool manufacturing Perlite manufacturing Phosphate rock processing Sand and gravel processing Stone quarrying and processing
9	Petroleum industry	Petroleum refining Crude oil distillation Catalytic cracking Hydrocracking Catalytic reformihg Polymerization, alkylation, and isomerization Hydrogen treating Chemical treating Physical treating Blending Katural gas processing
10	Wood processing	Chemical wood pulping Pulpboard Plywood veneer and layout operations Woodworking operations
11	Miscellaneous sources	Forest wildfires Fugitive dust sources Unpaved roads (dirt and gravel) Agricultural tilling Aggregate storage piles Heavy construction operartions

Table 3.13. Major Sources of Air Pollution

Point Sources	HC	NOx	CO	SOX	Part
Fossil-fueled electric power generating plants	x	o	x	o	0
Industrial boilers	o	o	o	o	o
Processing plants	o	x	x	x	o
Line sources					
Highways, roadways	o	o	o	x	x
Aircraft	x	o	o	x	x
Railroads	x	o	o	o	o
Area sources					
Indirect sources	o	o	o	x	x
Refineries	o	x	x	o	x
Residential tracts	x	o	x	o	x
Surface streeuaggrenated)	o	o	o	x	x

o - primary emphasis; x - secondary emphasis

Table 3.14. Air Pollutants from Various Electricity-Generating Technologies

Technology	Conversion Efficiency (percent)	Emissi.ons		
		NOx	SO2	C02
		(grams per lulowatt hour)		
High-sulfur coal-fired steam plant (without scrubbers)	36	4.3	21.1	889
High-sulfur coal-fired steam plant (with scrubbers)	36	4.3	2.1	889
Low-sulfur coal-fired fluidized bed plant	32	0.3	1.2	975
Oil-fired steam plant (uncontrolled)	33	1.4	1.6	794
Integrated gasification combined cycle plant (coal gasification)	38	0.2	0.3	747
Gas turbine combined cycle pl&t (current)	43	0.3	0	416
Gas turbine combined cycle plant (advanced)	55	0.03	0	331

Table 3.15. Geographical Scale of Impact of Ash, SO2, NOx, COa

Emission	Local	Rexional	National	Global
SO2	X	X	X	-
NOx	X	X	X	-
NH3	X	X	X	-
CO	X	X	X	-
HC	X	X	X	-
C02	-	-	-	X

X = sizeable impact ; XX = severe impact

Table 3.16. Composition of Dry Air in the Lower Troposphere (Free of Water Vapor)

Gas	Chemical Symbol	Concentration[a] (%)	Calculated residence time
Principal Gases			
Nitrogen	N2	78.0	continuous
Oxygen	O2	20.9	continuous
Argon	A	0.93	continuous
Carbon dioxide	CO2	0.032[b]	2-4 years
Trace Gases			
Permanent Gases (Nonreactive)			
Helium	He	5.2	about 2 million years
Neon	Ne	18	continuous
Krypton	Kr	1.1	continuous
Xenon	Xe	0.086	continuous
Reactive Gases			
Carbon monoxide	CO	0.1	0.5 years
Methane	CH4	1.4	4-7 years
Non-methane hydrocarbons	HC	01 x 10-3	?c
Nitrous oxide	N20	0.25	4 years
Nitric oxide	NO	0.2-2.0 x 10-3	5 days
Nitrogen dioxide	NO2	0.5-4 x 10-3	several days
Ammonia	NH3	6-20 x 10-3	7 days
Hydrogen sulfide	H2S	0.2 x 10-3	2 days
Sulphur dioxide	SO2	02 x 10-3	4 days
Chlorine	C12	3-15 x 10-4	several days
Iodine	I2	0.4-4 x 10-5	?
Hydrogen fluoride	HF	0.08-18 x 10-3	?
Hydrogen	H2	0.58	?
Ozone	03	0-0.05	about 60 days

a. This is the atmospheric background concentration, and not the concentrations found in polluted areas. When a range of concentrations is given, it indicates that these have been measured by different workers at different places; b. Minimum concentration of C02 measured away from centres of population. In population centres C02 concentrations vary from about 0.034% to 0.035%; c. Idttle is known about the residence time of the gas.

Dispersion of air pollutants emitted from industrial units is governed by topography of the air basin around the project site and meteorological conditions. Resultant change in air quality exert impacts as delineated in Tables 3.17 through 3.19 (Dhokalia, 1993).

Table 3.17. Effects of Suspended Particulates

Particulate Concentration (ugIm3)	Accompanying SO2	Duration or Averaging Time	Effect
1. Human Health			
750	715 ugh3	24 - hr. mean	Excess deaths and increase in illness may occur
300	630 ug/m3	24 - hr. mean	Chronic bronchitis patients likely to have a worsening of symptoms
200	250 ugh3	24 - hr. mean	Increased absence of industrial workers due to illness
100-130	120 ug/m3	Annual Mean	Children likely to experience increased incidence of some respiratory disease
80-100	Sulfation Rate 30 mg/cm2 mo	Annual Geo-metric Mean	Increased death rates for persons older than 50 years is likely
2. Visibility			
100-150			Where large smoke turbidity factors persist, iininiddle and high latitudes, sulight may be reduced by 1/3 ın summer and 213 in winter
150			For particles predominantloy in the 0.2 u to 1.0 u size range, and relative humidity less than 708, visibility can be reduced to as little as 5 miles
3. Materials			
60-180	Some	Annual Geo-metric Mean	In presence of SO2 and moisture, accelerated corrosion of zinc and steel may occur

Meteorological transport and diffusion processes, and chemical reactions in the atmosphere forp the fundamental link between pollutants emission rates and the resulting concentrations of pollutants in ambient air. Atmospheric dispersion modeling provides a scientific bases in the form of mathematical equations relating the release ı air pollutants into the atmosphere to ground level concentrations in ambient air. Such models have now been applied for several decades, and within this period of time, much efforts have been spent on making these models more reliable.

Table 3.18. Effects of Sulpher Dioxide Pollution

Concentrati on (ppm)	Duration or Averaging	Effect Time
1. Human Health		
0.052	24 - hr. mean	increased mortality may occur when particulates exceed 6 cohs* soiling index
0.25	24 - hr. mean	increased death and illness rate may occur whe smoke concentration is more than 760 Hglm3
0.21	24 - hr. inean	Patients with chronic lung disease may have worsened symptoms if smoke concentrations exceed 300 uglm3
0.19	24 - hr. mean	Increased mortality may occur if high particulate levels are present
0.11-0.19	24 - hr. inean	Increased hospital admissions for older persons with respiratory diseases
0.037-0.092	Annual mean	When accompanies by smoke concentrations greater than 185 ugIm3, increased frequency of lung and respiratory disease may occur
0.046	Annual mean	When accompanied by smoke concentrations greater than 100 uglm3, increased respiratory disease may occur in school children
2. Damage to Vegetation		
0.05-0.25	4 hr.	Moderate to severe injury to sensitive plants when 03 and NO are also present
0.03	Annual mean	Chronic damage and excessive leaf drop may occur. Some trees and shrubs show damage
3. Effect or Materials		
0.12	Ambient Mean	Corrosion rates for steel panels may be accelerated by 50%
0.09	Annual Mean	Some dyed fabrics fade
0.10	Ambient Mean	With 50% humidity and the presence of comparable particulate cocentration, visibility may be reduced to 5 miles

Table 3.19. Effects of Nitrogen Dioxide Pollution

Concentrati on (ppm)	Duration	Doses (ppm sec.)	Effect
1. Human Health			
A. Acute Effects due to Short-Tern1 Exposures			
0.12	-	-	Olfactory threshold
5	10 min	3,000	Measurable increase in airway resistance
15	2 hr	108,000	Tissue changes in lungs, heart, liver and kidney of monkeys

90	30 min	162,000	Pulmonary odeilla found 18 hr. after exposure-accompanies by a 50%-reduction in normal vital capacity
B. Effects due to Long-term Exposures			
0.062-0.109 Daily means	Continuous	37,000-60,000 per weak	Increased incidence of serious respiratory disease in fanlily groups found during a 6 month study
0.063-0.083	Continuous	38,000 50,000 per week	Frequency of acute bronchitis was increased among infants and school children observed in a 6 inonth study
5	6-24 hr/day for 3-12 nos.	9,720,000 - 158,000,000	Toxicological changes in mice reseinb ling those associated with huinan emphysema
2. Damage to Vegetation			
0.25	Continuous for 8 months	5,200,000	Leaf abscissioned decreased yield for naval oranges
0.50	Continuous for 35 days	1,512,000	Leaf abscission and chlorosis on citrus trees
1.0	1 day	86,000	Overt leaf injury to NOz sensitive plants
3. Other Effects			
A. Photochemical Oxidant (OX) Production			
0.04-0.16	-	-	Maximum daily 1 hr. average OX concentration of 0.1 ppm can be associated with these NO_2 levels in the presence of 0.3-1.4 ppm of nonmethane hydrocarbon (6-9 A.M. sunlight)

Air quality management strategies include, on one hand, ambient air quality standards (Table **3.20)** (Mishra & Mahadevan, 1991) and, on the other, promotion of dispersion of I>ollutants through tall stacks, and stipulation of concentration and mass based emission standards (Table 3.21) (MEF, 1993).

Table **3.20.** Summary of Ambient Air Quality Standards

Polluta nt	Avera- ging time	Country or Origin					
		USA	California	Japan	Canada	Czechos-lovakia	India*
SO2	Annual Av.	80	-	-	-	-	120 (I)
	24 Hr. Av.	365 @	131	120	180	150	80 (R)
	3 Hr. Av.	1300	-	-	-	-	30 (S)
	1 Hr. Av.	-	655	-	-	-	-
NO3	Annual Av.	100	-	-	-	-	120 (I)
	24 Hr. Av.	-	-	100	-	150	80 (R)
	1 Hr. Av.	-	475	-	-	-	30 (S)

CO	24Hr.Av.	-	-	1200	-	1000	5000(I)
	8 Hr. Av.	10000@	10000	25000	6000	-	2000(R)
	1 Hr. Av.	40000@	23000	-	16000	-	1000(S
Ozone (03)	1 Hr. Av.	235	180	-	-	-	-
Oxidants	24 Hr. Av.	-	-	40	-	100	-
(as 03)	1 Hr. Av.	160 @	-	100	-	-	TSPM
	Annual Av.	75 (GM)	-	-	160	-	500(I)
	24 Hr. Av.	260@..	-	-	-	-	200(R) 100(S
PMlO	Annual Av.	50 (AM)	30 (GM)	-	-	-	-
	24 Hr. Av.	150 (..)	50 (AM)	-	-	-	-
Hydro-carbons	3 Hr. Av. (6-9 AM)	160 (0.12 ppm)	-	-	-	-	-
			-	-	-	-	-
Lead	30 day Av.	-	1.5	-	-	-	-
	Calender Quarter	1.5	-	-	-	-	-
H2S	1 Hr. Av.	-	42	-	-	-	-
Vinyl chloride	24 Hr. Av.	26 0.01 ppm	-	-	-	-	-
			-	-	-	-	-
Sulphate	24 Hr. Av.	-	25	-	-	-	-

Notes

*I-Industrial, R-Residential, S-Sensitive : When monitored over the 12 months of a year with a frequency of not less than once in a week with a sampling time of either hours for any sample and analysed according to procedures specified by the CPCB the concentrations for the indicated pollutants shall be 95% of the time within the 1imits.prescribed indicated averaging period, not to exceed more than once per year

@ Currently the Hydrocarbon standard has been replaced by the Ozone standards in the USA.

Table 3.21. General mission Standards in India

I. Concentration Based Standards

S.No.	Parameter	Standard Concentration not to exceed (in mg/PTm3
1	Suspended particulate matter (SPM)	150
2	Fluoride	10
3	Abestos	4 fibrelcc
4	Mercury	0.2
5	Chlorine	15
6	Hydrochloric acid vapour and mist	35

7	Hydl-ogen sulphide	10
8	Sulphuric acid mist	50
9	Carbon monoxide	1%
10	Chloride	100
11	Lead	20
12	Sulphur dioxide	50

11. Equipment based standards

Control of sulphur dioxide emissions are achieved through dispersion. Minimum stack height limit is accordingly prescribed as below:

Parameter	Standard	
Sulphur dioxide	Stack-high limit in meters	
(i) Power generation capacity: - 500 MW and more - 2001210 MW and above to less than 500 MW - less than 2001210 MW	 275 220 H= 1 4 (Q)$^{0.3}$	
(ii) Steam generation capacity		
- Less than 2Tkr	Less than 8.5 MT	9
- 2 to 5 Tkr	8.5 to 21 MT	12
- 5 to 10 Tkr	21 to 42 MT	15
- 10 to 15 Tkr	42 to 64 MT	18
- 15 to 20 T/hr	64 to 104 MT	21
- 20 to 25 Tkr	104 to 105 MT	24
- 25 to 30 T/hr	105 to 126 MT	27
- More than 30 Tkr	More than 126 MT	30
	or using the formula H = 14 (Q) 0.3	

Note : H - physical height of the stack in metres
Q - emission rate of SO2 in kghr
MT - metric tones

III. Load/Mass-Based Standards

Sl. No.	Industry	Parameter	Standard
1	Fertilizers (Urea) Cominission prior to 1.1.82 Commissioned after 1.1.82	 SPM SPM	 2 kg/T of product 0.5 kg/T of product
2	Copper, lead and zinc smelting converter		4 kg/T of concentration (100%,) and produced
3	Nitric acid	Oxides of nitrogen	3 kg/T of weak acid before concentration produced
4	Sulphuric acid	Sulphur dioxide	4 kg/T of concerned (100%) acid produced

5	Coke over	Carbon monoxide 3 kg/T of coke produced
6	Oil refineries	

(a) For the oil refineries located in the area of Taj trapezium which include Mathura. Fatehpur Sikri, Agra and Bharatpur, the following standards shall be applicable

Process	Parameter	Standard
-Distillation (Atmospheric)	Sulphur dioxide	0.25 kg/MT of feed in this process
- Catalytic cracker	Sulphur dioxide	2.5 kgIMT of feed in this process
- Sulphur recovery unit	Sulphur dioxide	120 kg/MT of sulphur in the feed

(b) For the oil refineries located in the areas other than the Taj trapezium, the following stack height limits shall be applicable

IV. Dispersion through Tall Stacks

Plant type	Stack Height
1	For plants where the sulphur dioxide emission is estimated as Q(Kg/hr) the staclr height, H in metres is given by $H = 1\ 4\ (\ Q\ )^{0.3}$
2	For plants where the particulate matter emission is estimated as Q (tonneslhr.) the stack height, H in metres is given by H = 74(Q)OZ7
3	If by using the formula given in (1) or (2) above, the stack height arrived as, is more than 30 metres than this higher stack should be used
4	In no case should the height of the stack be less than 30 metres for plants given in the Schedule of the Air (Prvention and Control of Pollution) Act, 1981, and located in industrial areas of cities

Air quality models need to be used in EIAof industrial projects to :

- Concurrently implenient ambient air quality and emission standards through estimation of increase in levels of air pollutant concenentrations due to emissions from the industrial units
- Decide level of polltution control in industrial unit
- Select optimal stack height to avail benefit of assimilative capacity
- Design monitoring networks for baseline data collection as also post-project monitoring

USEPA developed a host of air quality models under the acronym UNAMAP for mainframe computers. These models are available through the National Technical Information Services (NTIS) of USA. R & D efforts over the years have reported developments on calibration of GPM .based models through correction coefficients for plume rise and preidcted ground level pollutant concentrations, and on development of 'front end devices' for user-friendly data entry in various UNAMAP models.

In India, the Bureau of Indian Standards (BIS) published guidelines in 1978 for ~nicrometereologica1s tudies in air pollution which interalia provide guidelines on air quality modelling. BIS modified these guidelines in 1993.

Use of UNAMAP packages poses problems similar to most of the 'canned' programmes. These include :

- Selection of appropriate model due to non-availability of specific models which can cater to needs of EL4 of industrial projects ,
- Expertise required in data entry and output interpretation
- Linking model outputs to graphics packages available commercially
- Limit to sensitivity analysis in terms of plume rise, wind direction and speed, and dispersion coefficients

Air quality models developed in the present research circumvent these limitations in user-friendly environment on PC-platform. The module on air quality prediction comprises three models :

- AQSCREEN for prediction of maximum ground level concentrations, distance to maximum ground level concentrations, and pollutant profiles for various stability conditions and wind speeds for single point source
- AQSHORT for prediction of short-term gorund level concentrations due to multiple point sources in complex terrain at user specified grid of (maximum) 480 receptors based on hourly metereological data. AQSHORT computes concentrations for user specified 4,8 or 24 hr averaging time as also sequential analysis
- AQLONG for prediction of long-term goru.ıd level conccntrations due to multiple point sources in complex terrain including valley situations at user specified grid of (maximum) 480 receptors based on stability rose (stabilitywind speed-direction- frequency) data. AQLONG computes concentrations assuming uniform lateral dispersion of pollutants.

3.5.1.2 Theoretical Background

The Gaussian plume model (GPM) is widely used for estimating dispersion of air pollutants over short distances from continuous point sources. The model is based on the premise that the plume has a Gaussian distribution and its concentration distribution at a given point down-wind can be expressed in terms of generalised Gaussian equation but with unequal dispersion coefficients in the horizontal and vertical directions.

Fundamental to the problem of atmospheric dispersion of pollutants is the concept of atmospheric stability which accounts for the phenomenon of atmospheric turbulence. The stability of the atmosphere is highly dependent upon the vertical distribution of temperature with height. Temperature of air in the troposphere normally decreases with height. The rate of change in temperature with elevation of a dry parcel of air when it is displaced adiabatically is known as dry adiabatic lapse rate (DALR). The rate at which the temperature of ambient air decreases with height is referred to as environmental lapse rate (ELR).

The temporal and spatial variations in ELR are quite significant in the lower layers of the atmosphere. A comparison of DALR to actual (environmental) lapse rate in the lower atmosphere indicates the stability of the atmosphere. Atmosphere is classified as unstable, neutral and stable when ELR DALR, ELR=DALR and ELR DALR, respecitvely. When the temperature increases with altitude, the lapse rate is negative and the atmospheric condition is termed as inversion. It is a condition of strong stability. The effect of an inversion is to reduce the vertical dispersion of pollutants and thus to increase the local concentration of pollutants.

The approach proposed by Pasquill (1961) for stability classification is the most widely used approach due to its simplicity and use of surface meteorologcal data. Despite the simplicity of this method, it displays the basic features of atmospheric turbulence. The instability of the atmospheric boundary layer increases with increasing insolation and decreasing wind speed, and this is what the Pasquill scheme prescribes, at least qualitatively. Pasquill defined six stability classes ranging from 'A' (extremely unstable) to 'F' (moderately stable). Surface wind, intensity of solar radiation and night-time sky cover have been identified as prime factors defining these stability categories (Table 3.22).

The most commonly used techniques to evaluate wind profile in lower layers of the atmosphere are logarithmic wind law and power law. As the logarithmic wind law is applicable only when the atmospheric conditions are neutral, it is a common practice to describe the wind profile with a power law which incorporates the concept of atmospheric stability. The typical values of wind extrapolation factor suggested by Irwin (1979) as a function of surface configuration and stability are presented in Table 3.23.

A particularly important aspect of the problem of atmospheric pollution is the determination of the height to which a plume will rise with initial flux of momentum and heat passing through the stack exit. Since maximum ground concentration is inversely proportional to the square of effective stack height, any increase in plume rise will significantly reduce maximum ground level concentrations. Plume rise, therefore, plays a vital role in air quality management.

Table 3.22. Pasquill Stability Classes

Surface wind speed (mlsc)	Day-time Isolation		Night-time conditions		
	Strong	Moderate	Slight	Thin overcast or 418 low clouds	3/8 cloudiness
<2	A	A-B	B		
2-3	A-B	B	C	E	E
3-5	B	B-C	C	D	E
5-6	C	C-D	D	D	D
>6	C	D	D	D	D

A- Extremely unstable
B - Moderately unstable
C - Slightly unstable
D = Neutral
E = slightly stable
F = Moderately stable

Table 3.23. Exponent for Wind Extrapolation

Paquill Stability Class	Urban	Rural
A	0.15	0.07
B	0.15	0.07
C	0.20	0.10
D	0.25	0.15
E	0.40	0.35
F	0.60	0.55

Plume rise is defined as the height between the top of the stack and the axis of the centroid of the pollutants distribution, and depends on :

- Diameter of the stack
- Exit velocity of effluent gases
- Temperature difference between the exit gases and ambient air
- Mean (horizontal) wind speed
- Atmospheric stability class
- Environmental lapse rate

The exact form of inter-relationship between these parameters depends on local circumstances such as terrain features, turbulence energetics etc. For this reason the problem of plume rise determination has remained particularly difficult to solve for practical applications. Carson and Moses (1969) tested 15 of the most commonly used plume rise fbrmulae against 711 observations of plume rise taken on various power plant sites (Sharma, 1993). Their conclusion was that there is no one formula which is outstanding in all respects, and more study is needed before a completely satisfactory method can be developed for predicting plume rise values. Briggs (1975) developed 'plume rise' equations based on a wide range of field data and a simple theoretical formulation which is in agreement with others. Briggs equations have been used for calculation of plume rise in the present research.

The dispersion parameters sy and sz are the standard deviations of the cross-wind concentration distributions in the lateral and vertical directions respectively, and are functions of down-wind distance X, surface roughness and turbulent structure of the atmosphere. As turbulence is fundamental to diffusion, it is a logical atmospheric variable and potentially most accurate as predictor of diffusion. Atmospheric turbulence is, however, difficult and expensive to measure and hence empirical turbulence typing schemes have been developed to handle practical atmospheric problems.

The majority of Gaussian models make use of dispersion parameters and classification methods proposed by Pasquill (1961), later modified by Gifford (1961) and Turner (1964). These dispersion parameters were deduced from tracer experiments with near-ground sources.

The turbulence typing schemes for tall stacks were presented by Singer and Smith (1966) based on several diffusion experiments performed at Brookhaven National Laboratory (BNL). Carpenter et ***al.*** (1971) developed curves of sy and sz for large coal burning power plants for the Tennessee Valley Authority (TVA).

Briggs (1973) critically examined P-G curves, BNL curves, and TVA curves and fitted the theoretical variation of sy and sz with distance predicted by Taylor's statistical theory. This led Briggs to deduce a new set of dispersion parameters which gave proper weightages to P-G curves as well as to BNL and TVA data. He proposed a series of interpolation formulae for the dispersion parameters in urban and open-country conditions. Briggs interpolation formulae for sy and sz are summarised in Table 3.24. These formulae give reasonable approximation of sy and sz over the distance range of 100 m to 10 km. Basis for the curves is presented in Table 3.25.

Table 3.24. Briggs Interpolation Formulae (100 < XO 10,000 m)

Pasquill Stability Type	Open-country conditions	
	σ y (m)	σZ (M)
A	0.22 x (1 + 0.0001x) -1/2	0.20 x
B	0.16 x (1 + 0.0001x) -1/2	0.12 x
C	0.11 x (1 + 0.0001x) -1/2	0.08 x (1 + 0.0002x) -1/2
D	0.08 x (1 + 0.0001x) -1/2	0.06 x (1 + 0.0015x) -1/2
E	0.06 x (1 + 0.0001x) -1/2	0.03 x (1 + 0.0003x) -1/2
F	0.04 x (1 + 0.0001x) -1/2	0.016 x (1 + 0.0003x) -1/2
Urban Conditions		
A-B	0.32 x (1 + 0.0004x) - 1/2	0.24 x (1 + 0.001x) -1/2
C	0.22 x (1 + 0.0004x) - 1/2	0.20 x
D	0.16 x (1 + 0.0004x) - 1/2	0.14 x (1 + 0.000x) - 1/2
A-F	0.11 x (1 + 0.0004x) - 1/2	0.08 x (1 + 0.0015x) - 1/2

Air quality modeling for multiple sources is based on the concept of superimposition which assumes that concentration from two or more sources is the sum of the concentrations that each source would produce in absence of all others. The concentrations at a receptor due to a series of point source is readily formulated by adding the concentrations of all sources.

Over extremely long periods (such as monthlseason/year) the wind direction is likely to touch on all points in the compass (Hanna, 1981). The horizontal crosswind distributibn in the plume. sy, is no longer important since wind-direction frequency distribution will be fairly smooth. In such cases sector model can be used, where it is assumed that there is no horizontal crosswind variation in concentration within the angular sector. The ground level concentrations at receptors are calculated according to Smith (1968) by assuming that the pollutant is uniformly distributed in the horizontal planes within the direction sector (-)and considering frequency of occurrence of wind in a particular direction, wind speed class, stability, and cumulative frequency of occurance.

Table 3.25. Basis and Scope of the P-G Curves

Crosswind spread:	
Source height	Any (within mixed layer)
Sampling time	3 min.
Basis for x = 0.1 to 1 km	Preliminary statistics of wind direction fluctuation for a surface roughness length of 3 cm.
Basis for x = 10 to 100 km	Extrapolation of short-range data in the light of limited special observations of tracer dispersion over level terrain of mixed roughness (implied roughness length of 30 cm.)
Vertical spread :	
Source height	Effectively zero (surface release of material). but offered as usable for any height in a mixed layer, in the absence of strong evidence to the contrary
Sampling time	Any. For elevated sources up to about 100 m. the limiting sampling time is roughly proportional to the height of the source; if the height of the source is above 100 m, the samplir~gti me is roughly 10 min.
Basis for x = 0.1 to 1 km	Properties of the wind profile over a surface of small roughness (roughness length of 3 cm), with guidance from dispersion studies, especially in regard to the effect of thermal stratification
Basis for x = 10 to 100 km	As for the crosswind spread, with guidance from early data on the properties of the vertical component of turbulence at heights throghout_the mixed layer

3.5.1.3 Mathematical Formulations

Mathematical Formulation for AQSCREEN and AQSHORT

The Gaussian equations based on coordinate scheme with the origin at the base of the stack, (x downwind from the source, y cross wind, and z vertical) can be expressed as :

For Stable Conditions or unlimited mixing,

$$Xp = Q * 1/u * g_1/ (\sqrt{2\pi}\ \sigma\ y) * g_2/(\sqrt{2}\ \pi\ \pi\ y) \quad (3.1)$$

where,

Xp = concentration (glm^3)

Q = emission rate (gls)

u = wind speed (m/s)
σy = horizontal dispersion coefficient (m)
σz = vertical dispersion coefficient (m)
g_1 = $\exp(-0.5\, y^2/\sigma y^2)$
g_2 = $\exp(-0.5(z-H)^2/\sigma z^2)\ \exp(-0.5(z+H)^2/\sigma z^2)$
y = cross wind distance (m)
z = receptor height (m)
H = effective stack height (m)

Under unstable and neutral conditions it is assumed that plume gets reflected for multiple number of times between ground, and the inversion layer and equation (3.1) gets modified to :

$$X_p = Q * 1/u * g_1/(\sqrt{2\pi}\,\sigma y) * I/L \text{ for } \sigma z > 1.6\text{ L} \qquad (3.2)$$

and

$$X_p = Q * 1/u * g_1/(\sqrt{2\pi}\,\sigma y) * g_2/(\sqrt{2\pi}\,\sigma z)$$

for sz 1.6 L and H & Z < L (3.3)

where,

L = mixing height (m)

$$g_3 = \sum_{N=S}^{N=5} \{\exp[-0.5(z-H+2NL)^2/ sz^2]+\exp[-0.5(z+H+2NL)^2/ sz^2]\} \qquad (3.4)$$

The model is based on following assumptions:

- an inert passive pollutant, i.e. no atmospheric chemical reactions and no gravity fall-out
- perfect reflection of the plume at the ground and inversion layer
- a non-zero wind speed so that diffusion in the x direction can be neglected in comparison with forward transport
- steady state conditions and homogeneous flow

Wind speed at stack top is estimated (Irwin, 1979) as:

$$\frac{Us}{U10} = \left(\frac{H}{10}\right)^p \qquad (3.5)$$

where,

Us = wind speed at top of stack, (mls)
U10 = wind speed at height of 10 m abovc ground, (mls)

H = physical stack height, (m)

Briggs (1975) plume rise equations have been used for calculation of effective stack height considering extrapolated wind speed at stack top as per terrain type (rural/urban) by using power lap equation and Irwins wind profile exponent.

Stack downwash has been considered through modification of the stack height as :

$$h' = h + 2 ((Vs/U(h)) - 1.5) d \text{ for } Vs< 1.5 U(h) \quad (3.6)$$

$$h' = h \quad \text{for } Vs = 1.5 U(h) \quad (3.7)$$

where,

h' = stack height corrected for downwash (m)

h = physical stack height (m.)

Vs = stack gas exit velocity (mls)

d = inside stack-top diameter (m)

h' is used throughout the remainder of the plume height computations.

For most plume rise situations, the value of the Briggs buoyancy flux parameter, F (m4 / m 3), is needed which is expressed as:

$$F = (g\ Vs\ d2)/ (4\ Ts) \quad (3.8)$$

Where,

F = buoyancy flux parameter (m4 I s 3)

g = acceleration due to gravity, 9.8192 mls 2

A 10= Ts - Ta

Ts = stack gas temperature (OK)

Ta = ambient temperature (OK)

Under unstable or neutral stability, for cases with stack gas temperature greater than or equal to ambient air temperature there is crossover between momentum and buoyancy phenomena which dominate plume rise. Hence, it must be decided whether plume is dominated by momentum or buoyancy. The cross over temperature difference (A T)c is determined for F less than 55 and F greater than or equal to 55. If the difference between stack gas temperature and ambient air temperature, Δ T exceeds or equals the (Δ T T)c, plume rise is assumed to be buoyancy dominated else as momentum dominated. Also, in situations where the stack gas temperature is less than the ambient air temperature, it is assumed that plume rise is dominated by momentum. (Δ T)c is calculated as:

For F < 55

$$(\Delta T)c = 0.0297\ Vs^{1/3}\ Ts/d^{2/3} \quad (3.9)$$

For F = 55

$$(\Delta T)c = 0.00575\ Vs2/3\ T \quad (3.10)$$

In the buoyancy dominated case, plume rise is estimated as:
For F < 55

$$H = h + 21.42\ F^{3/4}/U(h) \quad (3.11)$$

and

For F = > 55

$$H = h + 38.71\ F^{3/5}/U(h) \quad (3.12)$$

In the momentum dominated case, plume rise is estimated as:

$$H = h' + 3\ d\ Vs/U(h) \quad (3.13)$$

The distance to final plume rise (xf) in the case of buoyancy dominated case is etimated as :

For F < 55

$$xf = 0.049\ F^{5/8} \quad (3.14)$$

For F = > 55

$$xf = 0.119\ F^{2/5} \quad (3.15)$$

In the case of momentumdominated plume rise xf is taken as equal to 0:

Under stable conditions, for cases with stack gas temperature greater than or equal to ambient air temperature there is crossover between momentum and buoyancy phenomena which dominate plume rise. Hence, it must be decided whether plume is dominated by momentum or buoyancy. The cross over temperature difference (A T)c is determined for F less than 55 and F greater than or equal to 55. If the difference between stack gas temperature and ambient air temperature, AT exceeds or equals the (A Tb, plume rise is assumed to be buoyancy dominated else as momentum dominated. Also, in situations where the stack gas temperature is less than the ambient air temperature, it is assumed that plume rise is dominated by momentum. (A nc and is calculated as:

$$(\Delta T)c = 0.019582\ Vs\ Ts^{1/2} \quad (3.16)$$

where,

s = stability parameter

= 0.02 Wm for stability E

= 0.035 Wm for stability F

In the buoyancy dominated case, plume rise is estimated as:

$$H = h7 + 2.6\ (F/U(h)\ s)1/3 \quad (3.17)$$

and

$$H = h' + 4\ F1/4\ s^{-3/8} \text{ for calm conditions} \quad (3.18)$$

The lower of two values is taken as the final effective height. Simultaneous solution of equations 3.5.17 and 3.5.18 yields

$$U(h) = 0.2746\ F^{1/4}\ s^{1/8} \quad (3.19)$$

For wind speed less than or equal to this value Eqn. 3.5.17 is used for plume rise, else Eqn. 3.18 is used.

In the momentum dominated case, plume rise is estimated as:

$$H = h' + 1.5\ ((vs^2\ d^2\ T)/\ (4Ts\ U(h))^{1/3}\ s^{-1/6} \quad (3.20)$$

The distance to final plume rise (xi7 in the case of buoyancy dominated case is etimated as :

$$xf = 0.0020715\ U(h)\ s^{-1/2} \quad (3.21)$$

In the case of momentum dominated plume rise xf is taken as equal to 0.

For estimation of gradual plume rise and distance less than xf plume rise for any stability class is estimated as:

$$H = h'+ (160\ F^{1/3}\ x^{1/3})/U(h) \quad (3.22)$$

Dispersion coefficients in the present research are estimated using Pasquil stability classification and Pasquil-Gifford curves (Pasquil, 1976). Basis for the same is presented in Table 3.25.

P-G curves describing the verical spread of plumes are expressed as :

$$s\ z = a\ x^b \quad (3.23)$$

where,

x = downwind disance (km)

a, b = constants which are function of stability (Table 3.26)

Table 3.26. Constants for the vertical Dispersion parameter Equation

Stability Class	Distance (km)	a	b
A	<O. 1	122.80	0.9447
	0.1 - 0.15	158.08	1.0542
	0.15 - 0.2	170.22	1.0932
	0.2 - 0.25	159.62	1.1262
	0.3 - 0.4	217.41	1.2644
	0.4 - 0.5	258.89	1.4094
	>0.5	346.75	1.7283
B	<O. 2	90.673	0.93198
	0.2 - 0.4	98.483	0.98332
	>0.4	109.300	1.09710
C		61.141	0.91465
D	<0 . 3	34.459	0.86974
	0.3 - 1	32.093	0.81066
	1 - 3	32.093	0.64403
	3 - 10	33.504	0.60486
	10 - 30	36.650	0.56589
	>30	44.053	0.51179
E	<O. 1	24.260	0.83660
	0.1 - 0.3	23.331	0.81956
	0.3 - 1	21.628	0.75660
	1 - 2	21.628	0.63077
	2 - 4	22.534	0.57154
	4 - 10	24.703	0.50527
	10 - 20	26.970	0.46713
	20 - 40	35.420	0.37616
	>40	47.618	0.29592
F	<0.2	15.209	0.81558
	0.2 - 0.7	14.457	0.78407
	0.7 - 1	13.953	0.68465
	1 - 2	13.953	0.63227
	2 - 3	14.823	0.54503
	3 - 7	16.187	0.46490
	7 - 15	17.836	0.41507
	15 - 30	22.651	0.32681
	30 - 60	27.074	0.27436
	>60	34.219	0.21716

Lateral dispersion is estimated as:

$$sy = 1000 \times \text{Tan}(\Phi)/2.15 \qquad (3.24)$$

where,

x = downwind distance (km)

Φ = half angle from plume centrelinc to the edge of plume

= a ln(x) + b

a, b = constants which are function of stability (Table 3.27)

Table 3.27. Constants for lateral dispersion parameter equation

Stability class	a	B
A	-2.5334	24.167
B	-1.8096	18.333
C	-1,0857	12.500
D	-0.72382	8.3333
E	-0.54287	6.2500
F	-0.36191	4.1667

For strongly buoyant plumes, entrainment as plume ascends through the ambient air contributes to both vertical and horizontal spread. This induced dispersion is taken (Pasquill, 1976) as:

$$\sigma z\ \Phi = \Delta\ H/1.5 \qquad (3.25)$$

The effective dispersion is then determined as:

$$\sigma ze = (\sigma z^2 + \sigma z^2)^{0.5} \qquad (3.26)$$

Increased dispersion over urban areas is accounted in the present research through use of equations developed by Briggs (1973) presented in Table 3.24.

At the distance of final rise and beyond, the induced dispersion is constant based on the height of final plume rise. At the distance closer to the source, gradual plume rise is used to determine the induced dispersion.

Since in the initial growth of release, the plume is nearly symmetrical about its centreline, buoyancy induced dispersion in the horizontal direction, oycb, and effective lateral dispersion sye are equal to qzcb and sze respectively.

For multiple point source modelling the short-term concentration of a passive pollutant at grid point is calculated by the equation:

$$\mathbf{X\ (x,y) = \sum_{i-1}^{N} X\ (x,y)} \qquad \mathbf{(3.27)}$$

where,

X (x,y) = Total pollutant concentration at receptor located at co-ordinates x,y !ug/m3)

Xi = Pollutant concentration at receptor X (x,y) due to source i (pgfm3)

N = Total number of sources

Mathematical Fornulation for AQLONG

Pollutant concentration at the receptor in long-term mode is calculated as :

X (x,k) = 2 $*10^6$ * Q * F1/ E2 for $\sigma z <= 1.6$ L and H<L (3.28)

where,

$$F1 = \sum_{N=5}^{N=5} \{exp[-0.5(z-H+2NL)^2 / \%z^2] + exp[-0.5(z+H+2NL)^2 / \%z^2]\}$$

$$F2 = (2\pi)^{0.5} * U(k) * \sigma z * (2\pi \; x / N)$$

N = number of wind directions considered = 16

Hence F2 = 0.9843 * U(k) * %z * x

$$X(x,k) = 0.3927 * 10^6 \; Q / U \; L \; H \; x$$

for $\sigma z > 1.6$ L (3.29)

Total concentration at the recptor due to specific source is calculated as :

$$xrs = \sum_{d=16}^{d=16} \sum_{n=1}^{n=6} \sum_{s=1}^{s=6} Fdns \; X \; dns \qquad (3.30)$$

where,

Xrs = long-term conc, for specific receptor,r and due to source, s

F'dns = star data i.e. frequency of occurance for winddirection d, wind speed n, and stability class s during the intreasted period

Xdns = ground level concentration at receptor x

The concentration Xr at specific receptor due to all sources for a given pollutant is -

$$Xr = \sum Xrs \qquad (3.31)$$

In short-term mode AQLONG can be used assuming that a particular meteorological condition exists for any 6 hrs of a 24 hr period. This is accomplished within the model by reducing the input emission rate quarter (i.e. 1/4). This gives 24 hr average concentration assuming ***6*** hr steady-state meteorollogical condition.

For uneven terrain condition, plume rise and dispersion parameter in z direction are calculated using the same formulae as short-term model.

In the case of valley option, for stable conditions, the model assumes that the plume height above stack base remains constant after final plume rise. Thus, as terrain rises the plume approaches the elevated suface, and in effect the plume height decreases. Further, it is assumed that the plume centreline maintains a minimum 10 m. distance from the elevated terrain. For unstable and neutral atmospheric conditions plume is assumed to maintain a constant height above the terrain. The plume parallels the terrain feature by adjusting the effective height realtive to stack base, i.e. the situation is similar to flat plane.

3.5.1.4 Computational Representation

The computational procedure and data requirements are incremental in the models AQSCREEN, AQSHORT and AQLONG.

AQSCREEN model requires input from user on stack emission ' rate, exit velocity, stack top diameter, exit temperature, ambient temperature and terrain type (urbanlrural). For each stability class wind speeds to be considered are ibuilt in the model. Extrapolated wind speed and plume rise is computed by the model to estimate ground level concentration at various distances in downwind direction. Maximum ground level concentration and distance to maximum ground level concentration are then calculated in an iterative mode. AQSCREEN also reports the ground level concentration at various downwind distances to develop concentration profiles.

AQSHORT and AQLONG are multiple sources model. These models need definition of sources in a cartesian grid. Two orthogonal axes are used with the x-axis oriented in the east-west direction and y axis in North-South direction. Receptors are assumed to be located along the sixteen wind directions at spacing specified by the user. These models can handle maximum of 480 receptors and 50 sources.

In addition to source emission data, AQSHORT requires hourly data for wind speed, direction, stability class, temperature and mixing height. In the complex terrain option, user has also to provide terrain elevation at each receptor. Computational procedure involves estimation of pollutant concentrat~on due to each source at each receptor for each hour and arithmct~ca ddition of impact of multiple sources at receptor for the specified hour. The model reports concentration at receptors due to all sources for 8 and 24 hours averaging time. Concurrently, a three column oiltput file is generated for use in standard graphics packages. AQSHORT has inbuilt option of sequential analysis to estimate maximum 24 hour concentration at various receptors as required by the regulatory agencies.

AQLONG model requires metereological data in the form of stability rose (STAR), i.e. frequency of occurrence in each wind-speed class (maximum 6 classes, defined by user) in each stability class and each direction in addition to source and receptor data. Computational procedure involves estimation of pollutant concentration due to each source at each receptor for each hour and arithmetic addition of impact of multiple sources at receptor for the specified hour. The model reports average concentration at receptors for the period for which STAR data is provided by the user and, concurrently a three column output file is generated for use in standard graphics packages. User can operate AQLONG in short-term mode by inputing frequency equal to one for a particular combination of wind-speed and stability. It is assumed in such condition that the meteriological condition persists for a maximum of 6 hours in a day and accordingly the output concentrations are scaled down by a factor of 4. This mode of operation is extremely useful in designing monitoring network in multiple point source situation where the distance between the stacks is more than 150 m.

**The overall computational procedure adopted in AQSCREEN, AQSHORT and AQLONG comprises following steps :

- Read source data such as co-ordinates (x,y,z), emission rate, temperature, velocity, ambient temperature, option for plume rise computation (fixedlgraduavvarying)
- Read number of receptors in each of the 16 directions (maximum 30 in each direction), receptor spacing, terrain type (urbanlrural), ground elevation at receptors (for complex terrain)
- Read metereological data hourly1STAR
- Compute plume rise and distance to maximum plume rise using Briggs formulae using extrapolated wind speed
- Compute downwind and crosswind distance for each receptor
- Compute horizontal and vertical dispersion parameters (sy and sz) at the receptor

- Compute pollutant concentration at each receptor due to each source using appropriate form of GPM equation
- Compute total pollution concentration for the specified averaging time (8124 hr or long-term)

Flowchart for prediction of impact on air quality using STAR data is presented in Figure 3.6.

3.5.2 Prediction of Impacts on Water Quality

3.5.2.1 Problem Identification

Serious concern has been raised in India since early 1980s about deteriorating water quality in the river basins as a result of wastewater discharges from human settlements and industrial developments along the river banks.

Strategy for management of river water quality in India comprises use based classification of rivers (Table 3.28) on one hand and regulatory mechanism in the form of concentration and mass based effluent standards specified under the Environment Protection Act (1986) presented in Table 3.29 and 3.30. Implementation of this strategy warrants use of water quality models to establish cause-effect relationsbips.

A water quality model relates the impact of wastewater discharges into a river to the changes in its water quality with the help of mathematical relationships. The basic idea in modelling of water quality is to have a rational basis while making decisions on water quality management. Water quality models need to be used in EIA of industrial projects to :

- Predict impact of wastewater discharge on river water quality
- Decide the level of wastewater treatment
- Design of post-project water quality monitoring programme
- Set limits to abstraction of water with a view to maintaining minimum flow in rivers from assimilative capacity considerations

Sincc the the work carried out by Streeter and Phelps for Ohio river in the United States of America in 1920s for evolving relationships bettween DO and carbonaceous BOD, numerous water quality models of differing complexities have been pl,oposed and applied to many rivers. Such models include a large number of water quality parameters to describe the river water quality, such as DO, BOD, nitrogen'(N), phosphorus, algal concentrations etc. Models have also been proposed for dynamic conditions (e.g. in terms of fluctuations in flow) as well as for situations where river waters do not.mix completely with thewastewaters (e.g. in mixing zones). Thomann has provided a classical account of such developments (Thomann, 1987).

Table 3.28. Classification of Inland Surface Water

S.N.	Characteristics	A@	B@	C@	D@	E@
1	Dissolved oxygen, mg/l,Min	6	5	4	4	–
2	Biochemical oxygen demand, mg/l. Max	2	3	3	–	–
3	Total Coliform organisms,* MPN/100 ml, max	50	500	5000	–	–
4	Total Dissolved Solids,mg/l, Max	500	–	1500	–	2100
5	Chlorides(as Cl), mg/l, Max.	250	–	600	–	600
6	Colour, Hazen units, Max.	10	300	300	–	–
7	Sodium absorption ratio, Max.	–	–	–	–	26
8	Boron (as B) \, mg/l, Max.	–	–	–	–	2
9	Sulphates (as SO_4),mg/l, Max	400	–	400	–	1000
10	Nitrates (as NO_3),mg/l, Max.	20	–	50	–	–
11	Free Ammonia (as N), mg/l, Max.	–	–	–	1.2	–
12	Conductivity at 25°C, micromhos/, cm, Max.	–	–	–	10	2.25
13	pH value	6.5-8.5	6.5-8.5	6.5-8.5	6.5-8.5	6.5-8.0
14	Arsenic (as As),mg/l, Max.	0.05	0.2	0.2	–	–
15	Iron (as Fe) mg/l, Max.	0.3	–	50	–	–
16	Fluorides (as F), Mg/l, Max.	1.5	1.5	1.5	–	–
17	Lead (as Pb),mg/l, Max.	0.1	–	0.1	–	–
18	Copper (as Cu), mg/l, Max.	1.5	–	1.5	–	–
19	Zinc (as Zn), mg/l, Max.	1.5	–	15	–	–

* If the coliform count is found to be more than the prescribed tolerance limits, the criteria for coliforms shall be satisfied if not more than 20 percent of samples show more than the tolerance limits specified, and not more than 5 percent of samples show values more than 4 times the tolerance limits. Further, the fecal coliform should not be more than 20 percent of the coliform. Source: Indian Standard (IS : 229 - 1982).

@ A -Drinking water source without conventional treatment but after disinfection

B - Outdoor bathing (organised)

C - Drinking water source with conventional treatment tollowed by disinfection

D - Propogation of Wildlife. Fisheries

E - Irrigation, Industrial cooling, Controlled waste disposal

Table 3.29. General Standards for Discharge of Environmental Pollutants : Effluents

Sl. No.	Parameter	Standards			
		Inland surface water	Public Sewers	Land for irrigation	Marine coastal areas
1.	Colour and odour	See 6 of Annexure I	–	See 6 of Annexure I	See 6 of Annexure I
2.	Suspended solids mg/l, max	100	600	200	a. For process waste water -100 b. For cooling water effluent 10% above total suspended matter of influent
3.	Particle size of suspended solids	Shall pass 850 micron IS sieve-	–	–	a. Floatable solids, max. 3mm b. Settleable solids, max 850 microns
4	Dissolved solids (inorganic) mg/l Max.	2100	–	2100	–
5	pH value	5.5 to 9.0	5.5 to 9.0	5.5 to 9.0	5.5 to 9.0
6	Temperature	shall not exceed 5°C above the receiving water temperature	–	–	shall not exceed 5°C above the receiving water temperature
7	Oil and grease mg/l max.	10	20	10	20
8	Total residual chlorine mg/l max.	1.0	–	–	1.0
9	Ammoniacal nitrogen (as N), mg/l Max.	50	50	–	50
10	Total Kjeldahl nitrogen (as NH3); mg/l, max.	100	–	–	50
11	Free ammonia (as NH3) mg/l, max	5.0	–	–	5.0
12.	Biochemical oxygen demand (5 days at 20°C), max.	30	350	100	100
13.	Chemical Oxygen demand, mg/1 max.	250	–	–	250

Table 3.30. Wastewater Generation Standards

Sl. No.	Industry	Quantum
1	Integrated Iorn & Steel	16 M^3/ton of finished steel
2	Sugar	0.4 M^3/ton of cane crushed
3	Pulp & paper industries	
	a. Larger pulp & paper	
	i. Pulp & paper	175 M^3/tone of paper produced
	ii. Rayon grade pulp	150 M^3/ton of paper
	b. Small pulp & paper	
	i. Agro-residue based	150 M^3/ton of paper produced
	ii. Waste paper based	50 M^3/tone of paper produced
4	Fermentation industries	
	a. Maltry	3.5 M^3/ton of grain produced
	b. Brewery	0.25 M^3/KL of beer produced
	c. Distillery	12 M^3/KL of alcohol produced
5.	Caustic soda	
	a. Membrane cell process	1 M^3/ton of caustic soda produced excluding cooling tower blowdown
	b. Mercury cell process	4 M^3/ton of caustic soda produced (mercury bearing), 10% below down permitted for cooling tower
6.	Textile industries	
	Man-made fibre	
	i. Nylon & polyster	120 M^3/ton of fiber produced
	ii. Viscose rayon	150 M^3/ton of product
7	Tanneries	28 M^3/tone of raw hide
8	Starch, Glucose and related products	8 M^3/ton of maize crushed
9.	Dairy	3 M^3/KL of milk
10.	Natural rubber processing industry	4 M^3/ton of rubber
11.	Fertilizer	
	a. straight nitrogenous fertilizer	5 M^3/ton of urea or equivalent produced
	b. Straight phosphoric fertilizer (SSP & TSP) excluding manufacture of any acid	0.5 M^3/ton of SSP/TSP
	c. Complex fertilizer	Standards of nitrogenous and phosphatic fertilizers are applicable depending on the primary product

The choice of EIA team in the use of water quality models for prediction of impact of industrial developments is limited either to the classical Streeter Phelps equation for DO-BOD, or to use complex models such as QU.4L-2E.

While the first option appears convincing and adequate for EIA, there are serious limitations which include:

- Streeter-Phelps model in it's original form is applicable to single point source of pollution
- As the reaction rate coefficicnt for reaeration is a function of depth and velocity, estimation of the same in real life situations warrants iterative procedure
- Formulae for estimation of reaeration rate are empirical in nature and need correction for the stream stretch under study
- Deoxygenation rates based on bottle experiments can be significantly different from those in streams (Wright and McDonell, 1979)
- Oxygen demand due to ammonia nitrogen can be substantial in the case of discharge of nitrogen bearing wastewaters.

Models such as QUAL-2E, often regarded as comprehensive, require significant computational resources and demand considerable expertise for application. At its present state, QUAG2E requires some degree of modelling sophistication and expertise on the part of a user (Banvell et *al.*, 1987). The user must supply more than 100 individual inputs, some of which require considerable judgment in estimation.

This is illustrated through the range of coefficicnt values (Table 3.31) that user needs to select from for use of model. Further, field application of such models require substantial data inputs which are often not feasible to provide both on the grounds of constraints on financial resources as well as due to limitations of instrumentation and skilled manpower. Also, the user is often tempted to overlook the basic step involving selection of site specific water quality parameters for modelling.

These inherent difficulties in complex water quality models are especially relevant to the Indian situations. It is considered prudent to focus on the constraints on the data collection while choosing the rxodel. Practical utilization of the model should always be the overriding concern while visualizing the model dovclopment. The software for water quality model should be developed in affordable and PC friendly environment and should facilitate:

Table 3.31. Typical Ranges for QUAL2E Reaction Coefficients

Variable	Description	Units	Range of Values	Variable by Reach	Temperature Dependent
$\alpha 0$	Ratio of chlorophyll-a to algal biomass	$\frac{ug-Chla}{mgA}$	10-100	No	No
$\alpha 1$	Fraction of algal biomass that is nitrogen	$\frac{mg-N}{mgA}$	0.07-0.09	No	No
$\alpha 2$	Fraction of algal biomass that is phosphorus	$\frac{mg-p}{mgA}$	0.01-0.02	No	No
$\alpha 3$	O2 production per unit of algal growth	$\frac{mg-0}{mgA}$	1.4-1.8	No	No
$\alpha 4$	O2 uptake per unit of algae respired	$\frac{mg-O}{mgA}$	1.6-2.3	No	No
$\alpha 5$	O2 uptake per unit of NO2 oxidation	$\frac{mg-O}{mg-N}$	3-4.0	No	No
$\alpha 6$	O2 uptake per unit of NO2 oxidation	$\frac{mg-O}{mg-N}$	1.0-1.14	No	No
μ max	Maximum algal growth rate	day-1	1.0-3.0	No	No
ρ	Algal respiration rate	day-1	0.05-0.5	No	No
KL	Michaelis-Menton half-saturation constant for light (Option 1)	mtu/ft2-min	0.02-0.10	No	No

Table 3.31 contd...

KN	Michaelis-Mention half-saturation constant for nitrogen	mg-N/L	0.01-0.30	No	No
KP	Michaelis-Menton half-saturation constant for phosphorus	mng-P/L	.001-0.05	No	No
λ	Non-algal light extinction coefficientft	ft-1	Variable	No	No
λ	Linear algal self-shading coefficient	$\frac{1/ft}{ug}$ Chla/L	0.002-0.02	No	No
λ	Nonlinear algal self-shading coefficient	$\frac{1/ft}{(\lambda g\ Chla/L)^{2/3}}$	0.0165 (Riley)	No	No
PN	Algal preference factor for ammonia	–	0.0-0	No	No
σ1	Algal settling rate	ft/day	0.5-6.0	Yes	Yes
σ2	Benthos source rate for dissolved phosphorus	$\frac{mg-P}{ft^{2-}\ day}$	Variable	Yes	Yes
σ3	Benthos source rate for ammonia nitrogen	$\frac{mg-O}{ft^{2-}\ day}$	Variable	Yes	Yes
σ4	Organic nitrogen settling rate	day-1	0.001-0.1	Yes	Yes
σ5	Organic phosphorus settling rate	day-1	0.001-0.1	Yes	Yes
σ6	Arbitrary non-conservative settling rate	day-1	Variable		

Table 3.31 contd...

σ7	Benthal source rate for arbitrary non-conservative settling rate	$\frac{mg - ANC}{ft^{2-} day}$	Variable		
K1	Carbonaceous deoxygeneration rate constant	day-1	0.02-3.4	Yes	Yes
K2	Reaeration rate constant	day-1	0.0-100	Yes	Yes
K3	Rate of loss of BOD due to settling	day-1	-0.36-0.36	Yes	Yes
KR	Benthic oxygen uptake	$\frac{mg - 0}{ft2 - day}$	Variable	Yes	Yes
K5	Coliform die-off rate	day-1	0.05-4.0	Yes	Yes
K6	Arbitrary non-conservative decay coefficient	day-1	Variable	Yes	Yes
ß1	Rate constant for the biological oxidation of NH3 to NO3	day-1	0.10-1.00	Yes	Yes
ß2	Rate constant for the biological oxidation of NO2 to NO3	day-1	0.20-2.0	Yes	Yes
ß3	Rate constant for the hydrolysis of organic N to ammonia	day-1	0.02-0.4	Yes	Yes
α4	Rate constant for the decay of organic-P to dissolved-P	day-1	0.01-0.7	Yes	Yes

- Minimization of field data requirements
- Selection of coefficients based on literature, and calibration using field data
- Evaluation of impact of multiple point sources on water quality.

3.5.2.2 Theoretical Background

One of the most important considerations in predicting the impact of wastewater discharges on stream water quality is the ability of stream to maintain adequate dissolved oxygen concentration. Several combinations of variables can be significant in dissolved oxygen analysis in streams and rivers as shown in Fig. 3.7 (Pawlow, 1983).

The transformation processes which can be incorprated in an all- inclusive 1-D river water quality modelling are (Dobbins, 1964, Gromiec ***et al.*** 1983) :

- Bacterial oxidation of organic matter resulting in the exertion of carbonaceous and nitrogenous BOD
- Atmospheric reaeration in streams
- Reduction of BOD by sedimentation andlor adsorption
- Increase in BOD from scoured bottom deposits or from the diffusion of partly decomposed organic products from benthic layer into water above
- Increase in BOD along the stream from surface runoffs
- Removal of oxygen from the water by diffusion into the benthic layer to satisfy the oxygen demand in the aerobic zone of this layer
- Removal of oxygen from water by purging action of gases rising from the benthic layer
- Supply of oxygen by respiration of plankton and fixed plants Continuos redistribution of both BOD and DO by longitudinal dispersion
- Variation of deoxygenation coefficient with time, particularly at the onset of .nitrification
- Changes in channel configuration that alter the characteristics of surface turbulance and consequently the rate of transfer of oxygen from atmosphere

- Effect of suspended and dissolved substances upon the rate of transfer of oxygen from surface into main body of stream
- Diurnal variation in oxygen content, BOD, temperature and flow rate of influent discharges

QUAL-BE employees option IV in Fig. 3.7 and incorporates interaction of nutrient cycles, algal production, benthic and carbonaceous oxygen demand, and atmospheric reaeration. In addition, the model also includes heat balance for computation of temperature, and mass balances for conservative minerals, coliform bacteria, and non-conservative constituents such as radioactive substances. The major problems in use of QUAL-BE for EIA of industrial projects are :

- QUAL-BE was evolved to devise strategy for water quality management through a waste allocation strategy. In absence of any in-built algorithm, the model requires calibration in trial and error mode to adjust **33** reaction rate coefficients used in the model.
- The situation after the praject implementation would be different from the baseline situation, selection of appropriate reaction rate coeEcients is a difficult task for use of QUAL-BE in EIA.
- Temptation to model all in-built processes is **QUAL-BE** which may not be relevant in river water quality mcdelling in Indian rivers for specific ELA situations (such as longitudinal disperion) and division of basin into less than requisite number of reaches and elements to minimize data inputs results in numerical dispersion due to finite difference algorithm of QUAL-2E.

These problem can be resolved thorough :

- Selection of relevent parameters in modelling for which data can be collected in field with affordable manpower and financial inputs
- Use of\field data for calibration of model through in-built algorithm
- Providing over-riding choice to user for selection of coefficients as the nature of wastewater may change the baseline situation significantly

In view of this, the present research limits its scope to option II in Figure 3.7, viz, carbonaceous and nitrogenous BOD-DO interaction to develop module WQMODEL for EIA with further assumption that the settlable organic matter in the treated

of recyclelreuse options. BOD removal due to settling has also been neglected. The screening of modelling parameters also facilitates the use of analytical scheme for developing DO-BOD profiles.

3.5.2.3 Mathematical Formulations

The basic equation for one dimensional advection dispersion mass transport equation, which is numerically integrated over space and time for each water quality constituent, for any constituent C, can be written as :

$$\frac{\partial M}{\partial t} = \frac{\partial (A_x \text{ DsubL } \partial C/\partial}{Ax\partial x} \times dx = \frac{\partial (AX \text{ u } C)}{\partial x} dx + (AX\ dx)\ dC = \frac{S}{dt} \quad (3.32)$$

Where,

M = mass (M)

x = distance (L)

t = time (t)

C = concentration ($M\ L^{-3}$)

A_X = cross sectional area (L^2)

D_L = dispersion coefficient ($L^2\ T^{-1}$)

= mean velocity ($L\ T^{-1}$)

s = external sources or sinks ($M\ T^{-1}$)

In steady state condition the equation (3.6.1) reduces to :

$$\frac{\partial C}{\partial t} = \frac{\partial (A_x D_L \partial C/\partial x)}{A_x \partial x} - \frac{\partial (A_x \text{ u } C)}{A_x \partial x} + \frac{dC}{dt} + \frac{S}{V} \quad (3.33)$$

where, V = Ax.dx = incremental volume (L^3)

Dispersion is a predominant factor in tidal waters. In streams and rivers for continuous discharges and steady-state conditions, dispersion effects are usually insignificant (Pawlow, 1983). Hence the equation for application to streams and rivers further reduces to :

$$\frac{\partial C}{\partial t} = -\frac{\partial (A_x \text{ u } C)}{A_x \partial x} + \frac{dC}{dt} + \frac{S}{V} \quad (3.34)$$

The term dC/dt defines the changes that occur to individual constituents independent of advection, dispersion, and waste inputs, i.e. due to physical, chemical and biological reactions and interactions that occur in the stream. Examples of these changes are reaeration, algal respiration and photosynthesis, and colifornl die-off.

The basic relationship between DO-BOD proposed by Streeter and Phelps (1925) can be written as :

Change in BOD with time :

$$\frac{dBOD}{dt} = -K_i L \qquad (3.35)$$

where,

K1 = deoxygenation coefficient (day^{-1})

L = ultimate corbonaceous BOD, (mg/1)

The solution of equation 3.6.4 yields BOD at any time t as:

Lt = Lo exp (-Ki t) (3.36)

where,

Lt = ultimate carbonaceous BOD at time t

t = time of travel, days

Lo = initial ultimate carbonaceous BOD

Change in dissolved oxygen with time can be expressed as:

$$\frac{dD}{dt} = K_i L K_2 D \qquad (3.37)$$

where,

D = dissolved oxygen deficit at time t, (mg/l)

L = ultimate carbonaceous BOD at time t, (mg/1)

D = dissolved oxygen saturation deficit, (mg/l)

K1 = deoxygenation coefficient, (day^{-1})

K2 = reaeration coefficient, (day^{-1})

Solution of Eqn. 3.6.6 yields dissolved oxygen deficit at time t as :

$$Dt = \frac{K1 - K0}{K2 - K1} (\exp(-K1\ t) - \exp(-K2\ t)\ ] + D0 \exp(-K2\ t) \qquad (3.38)$$

where,

Dt = dissolved oxygen saturation deficit after time t (mg/l)

Lo = oxygen demand at an initial reference time (t=0) (mg/1)

Do = dissolved oxygen deficit at t=0 (mg/l)

t = traveling time in days

$$= \frac{x}{u}$$

x = distance (m)

u = mean velocity (m/d)

The basic assumptions in the Streeter-Phelps model are :

- Constant pollution load is discharged at a given single point along a stream having constant flow rate and uniform cross section
- Lateral and vertical concentrations of dissolved oxygen and BOD are uniform throughout any cross section
- Deoxygenation and reaeration are first-order reactions
- Reaction rates do not change with time
- Net change of dissolved oxygen deficit is a fiction only of deoxygenation and atmospheric reaeration by gas absorption through a liquid-air interface

Impact of nitrogen cycle on dissolved oxygen concentration can be expressed as (Miller *et al.,* 1979) :

$$\frac{dN\,1}{dt} = K_{11}\,N_1 \qquad (3.39)$$

$$\frac{dN\,1}{dt} = K_{22}\,N_2 + K_{12}\,N_1 \qquad (3.40)$$

$$\frac{dN\,1}{dt} = K_{33}\,N_3 + K_{23}\,N_2 \qquad (3.41)$$

$$\frac{dN\,1}{dt} = K_{44}\,N_4 + K_{34}\,N_3 \qquad (3.42)$$

where,

t = travel time (days)

N_1, N_2, N_3, and N_4 = concentrations of organic, ammonia, nitrite and nitrate nitrogen (mg/1)

K_{11}, K_{22}, K_{33} and K_{44} = deacy rates for organic, ammonia, nitrite and nitrate nitrogen (day")

K_{12}, K_{23}, K_{34} = forward reaction rates for organic, ammonia, and nitrite nitrogen (day^{-1})

Substituting the solution of the preceding equation into the next sequential equation and solving them results into concentrations of organic, ammonia, nitrite and nitrate nitrogen at time t (Miller *et al.,* 1979).

A simplified approach for incorporating ~itrogenous oxygen demand in Streeter-Phelps model expresses change in nitrogenous oxygen demand as (Bathala *et al.,* 1979, Constable & Mcbean, 1979, and Gromiec, 1983):

$$\frac{dBOD}{dt} = -K_n N \quad (3.43)$$

where,

N = initial ultimate nitrogenous BOD remaining at time t (mg/l)

K_n = deoxygenation coefficient for nitrogenous BOD (mg/l)

Total BOD at time t can be expressed as :

$$BOD_t = L_o \exp(-K_1 t) + N_o \exp(-K_n t) \quad (3.44)$$

where,

N_o = initial ultimate nitrogenous BOD, mg/l

Incorporating nitrogenous oxygen demand, the dissolved oxygen deficit at time t can be expressed as :

$$Dt = \left(\frac{K_1 L_0}{K_2 - K_1}\right) [\exp(-K_1 t)(-K_2 t)] + \left(\frac{K_1 L_0}{K_2 - K_1}\right)$$

$[\exp(-K_n t) \exp(-K_2 t)] + D_0 \exp(-K_2 t)$(3.45)

where,

D_t = dissolved oxygen saturation deficit after time t (mg/l)

L_o = ultimate carbonaceous BOD at an initial reference time (t=0) (mg/l)

N_o = ultimate nitrogenous BOD at an initial reference time (t=0) (mg/l)

K_1 = deoxygenation coefficient for carbonaceous BOD, (day^{-1})

K_n = deoxygenation coefficient for nitrogenous BOD, (day^{-1})

K_2 = reaeration coefficient, (day^{-1})

D_o = dissolved oxygen deficit at t=0, (mg/l)

t = travailing time (days)

$= \frac{x}{u}$

x = distance (m)

u = mean velocity (m/d)

Nitrogenous oxygen demand (NOD or NBOD) is exerted by both organic nitrogen and ammonia present in the wastewater. However, organic nitrogen needs to be first hydrolized to ammonia before its oxygen demand is exerted. Time and space lag in resultant oxygen profile, due to this sequential reaction, may be significant (Pawlow, 1983). If the two species are combined then the apparent oxidation rate (Kn) will be lower than the actual first order oxidation rate of ammonia. The ratio of TKN to NH3-N affects the value of the overall oxidation rate. This ratio changes after treatment of wastewater and is significant in the selection of Kn.

Also, the oxidation **of** alnmcnia takes place in two phases, nitriteeand nitrate. In many situations the nitrite concentration in the stream, observad and calculated, is vezy low and tends to be uniform (Pawlow, 1983). Hence, for Lhe purpose of modeling, ammonia oxidation is considered as a single rection sequence, i.e. ammonia to nitrate. Thus Kn can he expressed as :

$$\frac{dNH_3}{dt}\ \frac{dNO_3}{dt} = -K_n\, C_{NH3} \qquad (3.46)$$

and

$$\frac{dD}{dt} = -4.57\, K_n\, C_{NH3} \qquad (3.47)$$

Kn ranges between 0.1 - 0.5 per day (Pawlow, 1983).

Thomas (1948) and Camp (1963) proposed inclusion of BOD removal due to settling and addition by sediment oxygen demand in the Streeter-Phelps model. Camp (1963) and Dobbins (1964) also included longitudinal dispersion of BOD for developing DO-BOD profiles.

Equations 3.44, 3.45 and 3.47 form the basis of the development of WQMODEL, a one dimensional steady state water quality model, in the present research. Hydraulically, WQMODEL is limited to simulation of time periods during which both the stream flow as well as input waste discharges are essentially constant. It is based on the assumption of complete mix condition, i.e. at the point of discharge, stream is assumed to be homogeneous across the width (laterally) and with depth (vertically). It allows for multiple waste discharges, withdrawals, tributary flows, and incremental inflow and outflow.

The stream to be modeled is first divided into various reaches. Within a reach, the hydrogeometric properties (e.g. bed slope, cross-section, roughness etc.) and rate constants (e.g. BOD decay rate, atmospheric reaeration rate etc.) are assumed to be constant. Mass can be added to or removed from the ,system via external sources or sinks such as biological transformations. Each reach is considered to be completely mixed. Equations 3.44 and 3.45 are used to estimate residual BOD and DO concentrations in the flow direction.

The average stream velocity is estimated using Mannings equation :

$$U = \frac{1}{n} R^{2/3} S^{1/2} \quad (3.48)$$

where,

U = average velocity (mls)

n = Manning's Coefficient

R = hydraulic radius (m)

= area/wetted perimeter

S = river bed slope

Typical values of n are available in literature (Pawalow, 1983).

In each reach, the depth of flow needs to be computed such that the velocity calculated by equation 3.48 satisfies the continuity equation. WQMODEL does this with adjustment using sequential search technique.

Actual stream cross section is approximated as trapezoidal in nature. Wetted area and perimeter are calculated as :

$$A = W \times D - 0.5\,(Sl \times D) \times D - 0.5\,(S2 \times D) \times D$$

$$P = W - S1 \times D - S2\,D + D\,(1+ Sl^2)^{0.5} + D\,(1+S2^2)^{0.5} \quad (3.49)$$

where,

A = wetted area (m2)

P = wetted perimeter (m)

U' = width of river at top of water level (m)

S1 = left side slope of the section (1 horizontal in S1 verticals)

S2 = right side slope of the section (1 horizontal in S2 verticals)

Dissolved oxygen saturation concentration is estimated out in the present research as **(APHA,** 1987) :

$$DO_s = 139.344 + \frac{(1.575 \times 10^5)}{T} - \frac{(6.642 \times 10^7)}{T^2} + \frac{(1.244 \times 10^{10})}{T^3} - \frac{(8.622 \times 10^{11})}{T^{12}} \quad (3.50)$$

where,

DOs = dissolved oxygen saturation concentration, (mg/l)

T = stream temperature after mixing, (OK)

The reaeration coefficient K2 is most often' expressed in terms of stream depth and velocity. WQMODEL estimates following the guidelines proposed by Covar is based on following equations:

If depth of flow <2 feet, then

$$K_2 = 21.714\ U^{0.67}\ d^{-1.85} \tag{3.51}$$

(Owens, Edward and Gibbs, 1964)

where,

U = average flow velocity (ft/sec)

d = average flow depth (ft)

If depth of flow > 2 feet and

If stream velocity > (1.5 + 0.0729 d - 2), then

then,

$$K_2 = 11.61\ U^{0.969}\ d^{-1.673} \tag{3.52}$$

(Churchill, 1962)

Else,

$$K_2 = 12.846\ U\ d \tag{3.53}$$

(O'Conner and Dobbins, 1958)

The temperature dependency of Kz is expressed in the form as:

$$K_2T = K_220\ (1.024)^{(T - 20)} \tag{3.54}$$

$K_2$20 is the reaeration coefficient at 20°C

Deoxygenation coefficient, Ki, depends on the type of waste, e.g. domestic waste, industrial waste etc. Even for a type of waste, e.g. domestic waste, Ki changes depending on the degree of waste treatment provided. For example, K1 may be 0.5lday for raw domestic waste, and the same could reduce to O.l/day for secondary treated domestic wastewater. Besides, K1 also depends upon a variety of factors such as water temperature, hydraulic factors and 'stream geometry. Ki also depends on the stream hydraulics. Moro commonly K1 is a function of flow rate, hydraulic radius as wit11 as stream depth. Wright and McDonne11 (1979) utilized data from 36 stream reaches in the US to derive the expression :

$$K_1 = 10.34\ Q^{-0.486} \qquad (3.55)$$

where,

Q = flow tft3isec)

Another expression was derived based on the saqle data using channel wetted perimeter as t;hc independent variable as f'ollows,

$$K_1 = 39.57\ P^{-0.839} \qquad (3.56)$$

where P = wetted perimeter (it)

Wright & McDonell (1972) observed that the deoxygenation coefficients vary between (1.38id to 4.25. Over thc range of flows 0.126 m^3/s to **214** m^3/S

For flow regimes greater than 22.4 m^3/s (800 cft) deoxygenation rates range narrowed to *0.08/d* to 0.30/d and were consistent with bottle rates.

In view of better correlation between flow and K1 (Wright and MCDonell, (1979), K1 is estimated as function of flow in the present research.

The temperature rlepc21idency of K1 is expressed in tile form as:

$$K_1T = K_120\ (1.047)^{(T-20)} \qquad (3.57)$$

where, K_120 is the deoxygerlation coefficient for carbonaceous BOD at 20°C

Field estimations of deoxygenation coefficients can be **on** higher side as it includes BOD rernoval by settling as well.

Deoxygenation coefficient for nitrogenous ROD is taken as user input in WQMODEL and then is corrected for temperature as:

$$K_nT = K_n20\ (1.047)^{(T-20)} \qquad (3.58)$$

where, K_n20 is the deoxygenation coefficient for nitrogenous BOD at 20°C

3.5.2.4 Computational Representation

WQMODEL allows modelling of DO-BOD for a single river system. River bifurcations and tributories need to be analysed seperately. WQMODEL can consider multiple point waste discharges, abstractions, tributory flows (as point loads), and outflows.

The first step involved in WQMODEL use is to divide the river strech into reaches. Beaning of a reach is termed as 'from node' and the location at the end is termed as 'to node'. WQMODEL defines following types of nodes in the representation of a stream system.

A 'Reference Node' is the node at the headwaters of the main branch or tributory of the river. This is the first node in WQMODEL computation.

A 'Waste Discharge Node' is a node where a point discharge of wastewater enters the river. Similarly, tributories are considered as waste discharge node,

An 'Abstraction Node' is a node where water is abstracted from river,

A node for 'Geometric Change' is a node where the stream hydraulics changes appreciably such as due to change in the cross sectional area or depth of flow.

A 'Tail Node' represents the end node of the river strech under study.

The nodes and reaches are numbered serially in the flow direction.

Model calibration step in WQMODEL involves the determination of the values of the coefficients used in the model to match the observed system behaviour. In the case of WQMODEL calibration mean estimation of K1, Kn and K2 so as to match the simulated DO-BOD concentrations with observed DO-BOD. WQMODEL uses a unique calibration strategy. Initially, the reaeration coefficients at each reach are estimated using the appropriate equation. Initial estimation of deoxygenation coefficient K_I are assumed in all the reaches based on equation **3.55.** The calibration routine operates on BOD first (i.e. adjustment of K1 and $\mathbf{K_n}$**)** and then works on fitting DO (i.e. adjustment of K2 based on calibrated values of Ki and Kn. Further, it moves sequentially from one monitoring station to another as per sequence of river flow. At the first monitoring station, a random search algorithm is used to find the optimal estimates of the model coefficients in the reaches upstream to the monitoring station in such a way that the deviation between the observed and the simulaed concentrations with respect to BODDO is minimum. This search for the best combination of paramter estimates is done with the constraint that the model coefficients are always within the limits. At the end of calibration, Root Mean Square Error (RMSE) is reported.

As the computation of coefficients and calibration routine result in diffent set of values, WQMODEL gives the user over-riding choice for modifying these values for prediction of impacts of industrial unit on the river water quality.

**The overall computational procedure in WQMODEL involves following steps:

- Read data on number of reaches
- For each reach read the stream stretch length, channel width at top, side slopes, bed slope, Mannings roughness coefficient, reaction coefficient for oxidation of ammonia nitrogen
- At node 1, reference node, read data on stream flow, temperature, initial CBOD, ammonia-nitrogen, and dissolved oxygen concentration
- At each node, if it is wastewater node, read data on wastewater flow, CBOD, ammonia-nitrogen, dissolved oxygen concentration and temperature
- For each reach, if it is abstraction node, read data on water abstraction rate
- In each reach, read data on distance of monitoring station from from node of reach, CBOD, ammonia-nitrogen, and dissolved oxygen concentration
- At each wastewaterlabstraction node perform flow balance and compute velocity and channel widthldepth after mixing1 abstraction using Newton Raphson method
- At each wastewaterlabstraction node perform mass balance to compute CBOD, ammonia nitrogen, dissolved oxygen and temperature after mixing
- Compute K_1 and K_2 for each through selection of appropriate equation and adjust the same for temperature correction
- Predict concentration of CBOD, ammonia nitrogen, dissolved oxygen at each monitoring station in the flow direction
- Calibrate sequentially at each monitoring station with respect to observed and simulated concentration of CBOD, ammonia nitrogen, dissolved oxygen to estimate K_1, K_n and K_2 while minimizing RMSE
- Seek user input for modifying values of K_1, K_n and K_2 for impact prediction

Develop DO-BOD profiles using modified reaction coefficients

Flowchart for WQMODEL is presented in Figure **3.8** (Please see at P. 123).

3.5.3 Prediction of Impacts on Noise

3.5.3.1 Problem Identification

The term noise is defined as unwanted and undesirable sound (Rau and Wooten, 1981). It is characterised by its loudness and frequency spectrum and is a function of human subjectivity. Physiological effects of noise exposure on human .beings are manifested in many forms. These are loss of hearing ability, temporary threshold shift, headache, nausea, loss of memry, hypertension and sleep-interference.

Table **3.32.** Noise produced by select equipment

S.N.	Equipment	Noise Level dB (A)
1	Air Compressor	95-104
2	Ball mill	91-96
3	Blower (in forging shop)	97
4	Chipping	110
5	Chiselling, pneumatic	118
6	Diesel generator, 110 KVA	95
7	Drill machine	100
8	Forging hammer, 2000 kg	130
9	Forging hammer, 5000 kg	135
10	Furnances	73-104
11	Jaw crusher	90-100
12	Pedestal grinder	92
13	Lathe machine	87
14	Milling machine	112
15	Oxy-acetelene cutting	96
16	Pumps	89-100
17	Pulverizer	95
18	Power press, 50 tonnes	109
19	Rivetting	95
20	Power Operated saw	108
21	Sieve shaker	87
22	Shot blasting	103
23	Steam turbine, 12500 kw	91

24	Tumblivg barrel	107

The major industrial sources of noise are illustrated in Table **3.32** (Madhusudanrao, **1991)** and noise levels in select industries are presented in Table **3.33** (Kapoor, **1993,** Sengar, **1993).**

Table 3.33. Noise levels in select industrial plants

S.N.	Industry type	Noise level dB (A)
1	Foundry	93-127
2	Textile	76-102
3	Chemical	89-100
4	Glass blowing shops	70-108
5	Peteroleum refinery	74-98
6	Vanaspati industry	76-87
7	Ammunation factory	85-101
8	Power Plant	90-100
9	Heavy vehicle factory	110-115

The strategy in India to minimize the impact of industrial noise sources comprises specifications for equipment manufactures (Table 3.34) (MEF 1993), and stipulation of ambient noise levels (Table 3.35 and Table 3.36) (MEF, 1989).

Table 3.34. Noise Emission Standards (at manufacturing stage)

S.N.	Type	Noise Level dB (A)
A	Automobiles	80
1	Motorcycle, Scooters & Three wheelers	80
2	Passanger cars	82
3	Passanger of commerical vehicle upto 4 MT	85
4	Passanger of commerical vehicle 4-12 MT	89
5	Passanger of commerical vehicle above 12 MT	91
B.	Comestic appliances and Construction equipment	
1	Window air conditioners 1-1.5 T	68
2	Air coolers	60
3	Refrigerators	46
4	Diesel generators	85-90
5	Conlpactors (rollers)	75
6	Front loaders	75
7	Concrete mixers	75
8	Cranes	75
9	Vibratiors	75
10	Saws	75

Table 3.35. Acceptable outdoor Noise Levels Is : 4954-1968

S.N.	Location	Noise Level. dBA
1	Rural	25-35
2	Sub-urban	30-40
3	Urban - Residential	35-45
4	Urban-Residual and business	35-45
5	City	40-50
6	Industrial Area	50-60

Table 3.36. Ambient Noise Standards

Area Code	Category of Area	Leq limits Days Time	dBA Night Time
A	Industrial area	75	70
B	Commercial area	65	55
C	Residential area	55	45
D	Silence zone	50	40

Despite the fact that the impacts of industrial noisc sources are not significant after a disance of **1** km due as the sound dissipates within a short distance from source, dies fast, and does not accumulate in the env;ronment like other pollutants, it is necessary to predict impa* of sources of noise in the proposed industrial unit on the sur nding arcas so as to implement the strategy delineated above.

The model for pred tion of noisc levels duc to industrial sources of noise developed **-1** this research, NQMODEL, estimates: Cumulative noise level due to a maximum of **10** noise sources for a receptor grid comprising **30** receptors in the North and East direction accounting for the height of sources and receptors above ground

Noise attenuation due to dispersion, physical barriers, and vegetation

Width of green belt to attenuate noise to the desired level

3.5.3.2 Theoretical Background

Three basic elements involved in the study of noise predictions are the source, the receiver, and the transmisssion path.

The sources of noise may be categorised as mobile and stationary sources; point, line and area sources; continuous, intermittent and implusive sources etc. Major sources of noise are transportatiop (Road, Rail & Air), industry, construction, community activities (Public address s em, crackers, etc), and home appliances.

The transmission path from source to receiver is characterized by the properties of the medium, i.e. air, and the presence of barriers, viz. walls, trees, vegetation etc. Noise is attenuated during transmission due to atmospheric absorption, scattered by natural and artificial barriers, and absorbed due to ground cover. The sound energy reaching a receiver may be due to a number of sources of different characteristics. The net energy reaching a receiver from number of sources may be computed by logarithmic addition of the noise level at the receiver due to all the source independently.

A proportion of the sound energy is converted into heat as it travels through the air. It is significant at high frequencies and over arge distances. The classical absorption is partly due to viscosity and heat conduction of air and mostly by molecular absorption. .The presence of water vapour in the air facilitates the excitation of resonant molecules by the sound wave. The absorption depends significantly upon the frequency and relative humidity, and to lesser a extent upon the temperature. Attenuation due to classical absorption can be related to atmospheric parametei.~ such as temperature, pressure and relative humidity. Walls or barriers locared in the transmission path between noise source and receptor reduces noise levels due to increase in the path length.

Vegetation affects the propagation of low frequency sound by ground absorption which can be enhanced by high porosity of the ground resulting from tree -roots and fallen leaves, etc. High frequency propagation is affected by scattering by tree trunks, branches and pr aAly by leaf absorption.

3.5.3.3 Mathematical Formulations

The sound level' at receptor due to indvidual sources is computed in the present research as (Rau and Wooten, 1981) :

$$Lp = Lw + DI - 20 \text{ Log } r - Ae - 11 \qquad (3.59)$$

where,

Lp = Sound pressure level at a receptor located at distance 'r' from source

Lw = sound power level of source (dB)

r = distance between source & receptor (m)

Ae = excess attenuation between source and receptor transmission pathway caused by environmental conditions

DI = directivity index of source in the direction 'O' (for hemispherical wave propagation DI = 3dB)

Total noise level at receptor is calculated as :

$$\text{Lp Total} = 10 \text{ Log } [\sum_{i=1}^{n} 10^{(Lpi/10)}] \quad (3.60)$$

where,

Lpi = noise level due to source 'i' (dB)

Excess Attenuation is computed as :

Excess attenuation due to air absorption :

$$\text{Ae (dB)} = 7.4 \times \frac{f^2 . r}{\phi} \times 10^{-8} \quad (3.61)$$

where,

f = frequency of sound (Hz)

r = distance between source and receptor (m)

f = relative humidity (%)

Excess attenuation due to barriers:

$$AeB = 7.351 + 7.498 \text{ x } N - 2.2968 \text{ x } (N)^2 + 0.3842 \text{ x } (N)^3 \quad (3.62)$$

where,

N = freznel number

$$= \frac{A + B + C}{\lambda}$$

where,

A = distance between source & top of barrier (m)

B = distance between receptor & top of barrier (m)

C = distance between source & receptor on ground (m)

λ= wavelength (m)

$$= \frac{C}{f}$$

where,

C = velocity of sound (m/s)

= 20.05 T

where,

T = absolute temperature (^0C)

Excess attenuation due to grass, shrubs and trees :

$$AeG = (0.18 \text{ Log } f - 0.31) \times r \quad (3.63)$$

where,

f = frequency of sound (Hz)

r = path length through shrubbery/grass (m)

Excess attenuation due to forests:

$$AeF = 0.01 \cdot f^{1/3} \cdot r \quad (3.64)$$

If sound power level at source is not known but sound pressure level is known, then the sound pressure level is computed as :

$$Lp_2 = Lp_1 - 20 \text{ Log } \frac{r2}{r1} - Ae_{1,2} \quad (3.65)$$

where,

Lp2 = sound pressure level at receptor 'R' (dB)

Lpi= sound pressure level at refereGce point (dB)

r l = distance of reference point from source 'Si' (m)

r2 = distance of receptor 'R' from source 'Si' (m)

Total hoise level at receptors as specified in terms of day and night sound level, Ldn, is computed as :

$$Ldn = 10 \text{ Log } ((15 (10^{Lptotal/10}) + 9(10^{(Lptotal + 10)/10})) \quad (3.66)$$

3.5.3.4 Computational Representation

NQMODEL assumes noise sources to be located wiithin the receptor -grid specified by the user in the first of orthogonal co-ordinate system, i.e. (0,O) at the extreme left corner at bottom. User specifies source co-ordinates (x,y,z) and recptor spacing to define the computational frame. User has option of modifying the receptor spacing to derive better spatial resolution.

Barriers (maximum 10 nos) in between sources and receptors need to be defined as (x,y) co-ordinates at the begining and end of each barrier. User also provides average height of the barriers.

Grass, shrubs and forests need to be defined as polygons of a maximum of five sides and user provides (x,y) co-ordinates at each corner of the polygon.

The overall computational procedure adopted in NQMODEL comprises:

- Read basic data about number of sources, &ed receptors, recptor spacing, ambient pressure and temperature, and background noise level
- Read source data, **viz.** sound pressure level, frequency and location
- Read co-ordinates (x,y,z) for fixedlsensitive receptors and height for individual grid receptor
- Compute noise level at each receptor due to each source accounting for attenuation due to air absorbtion, barriers, vegetation and forest
- Comkte total noise level at each receptor
- Compute day and night noise level at each receptor

Flowchart for NQMODEL is presented in Figure **3.9** (Please see at P. 124).

3.5.3.5 Biological Environment

In absence of established tools prediction of impact of industrial project on biological environment, it is proposed in present research that indicators of biological environment be estimated for baseline scenario. Present research proposed that the expert judgement should be sought to predict changes in these indicators in qualitative terms.

3.5.3.6 Socio-Economic Environment

Major impact of industrial project on socio-economic environment, would be due to exmployment opportunities made available by the project. ***All*** other impacts on socio-economic environment would be triggered-off due to the workforce imigration. Present research work proposes that the expert judgement should be sought to predict impact on socio-economic indicators, which define quality of life, in qualitative terms.

3.6 IMPACT 'EVALUATION

The basic objective of impact evaluation exercise is to cpmmunicate to decision makers the net result of proposed project on environment. This warrants lumping of parameters indicaing status of environment with and without the project as also devising appropriate environmental management plan to minimize adverse impacts.

Methodology for impact evaluation developed in the present research draws upon the environmental evaluation system developed by the Battelle laboratory (Whitman, 1971), Leopold Matrix as modified by Lohani and Tahn (1982) for selection of parameters for impact evaluation as a logical outcome of impact identification exercise. Accordingly, the parameters have been divided in four categories, viz.

- Ecology
- Environmental Pollution
- Aesthetics
- Human interest

These categories have been further divided into eightcomponents which group in all 51 paramters as shown in Table 3.37.

Table **3.37.** Parameters for Impact Evaluation of Industrial Projects

Category	Component	Parameter
Ecology	Aquatic ecology	River flow Aquatic natural vegetation Phytoplankton diversity Zooplankton diversity Benthal flora and fauna Algal growth Fish yield
	Terresttial ecology	Green cover Crop yield Rare & endangered plant species Rare & endangerd animal species Vegetation diversity Fauna diversity
Environmental Pollution	Air	Suspended particular matter Oxides of sulphur Oxides of nitrogen Carbon monoxide Total hydrocarbons Photochemical oxidants
	Water	Dissolved oxygen Biochemical oxygen demand Temperatures Faecal coliform Total phosphate Ammonia nitrogen Turbidity Total dissolved

	Land	Exchangeable sodiuln & Water extractable metals, Soil permiability, Soil leachate quality
	Noise	Noise Level - sparce Noise Level - frequent Noise Level - continuous
Aesthetics	Aeshthetics	Landscape Appearnace of waterbody Odour nuisance Fly nuisance Visibility Composite effect
Human Interest	Socio-economics	Housing facilities Employment opportunities Health facilities Sanitation facilities Educational facilities Population desnity Public Transport facilities Drinking Water supply Power supply Perishable food prices Land Price

The major problem in application of Battelle Environmental Evaluation System (BEES) to evaluation of industrial projects relate to quantification of impacts and subjectivity in assigning weights.

Prediction models for air, water and noise environments developed in the present research provide quantitative estimates for environmental pollution. In the absence of established techniques for prediction of impacts on ecology and socio-economic environment, expert judgement in the form suggested by Leopold (importancelmagnitude values) needs to be sought. Further, in the view of spatial dimension of impacts, impact-zones need to be defined. Parameter measurement units and impact zones developed in this research are presented in Table 3.38.

Table 3.38. Parameter Measurement for Impact Evaluation

S.N.	Parameter	Measurement Unit	Measurement/sigpificance
	Category : Ecology Component : Aquatic ecology Impact zone : Wastewater receiving waterbody		
1	River flow	Irnpact - 1 to +1	Withdrawal of river water and disch- arge of wastewater in liver would change river flow. Impact on river flow would depend upon industrial water require

			ments. Reduction in river flow would adversely affect aquatic ecosystem. In absence of prediction technique impact measurement would be subjective in which - 1 corresponds to highly adverse and +1 to highly beneficial impact.
2	Aquatic natural vegetation	Impact -1 to +1	Dischare of industrial wastewater would affect aquatic natural vegetation based on presence of nutrients, organic matter, heavy metals and toxic substances. In absence of prediction technique impact measurement would be subjective in which -1 corresponds to highly adverse and +I to highly beneficial impact.
3	Phytoplankton diversity	Impact -1 to +1	Discharge of industrial wastewater would affect phytoplankton diversity based on presence of nutrients, organic matter, heavy metals and toxic substances. In absence of prediction technique impact measurement would be subjective in which -1 corresponds to highly adverse and +1 to highly beneficial impact.
4	Zooplankton diversity	Impact -1 to +1	Discharge of industrial wastewater would zooplantion diversity based on presence of nutrients, organic matter, heavy metals and toxic substances. In absence of prediction technique impact measurement would be subjective in which -1 corresponds to highly adverse and +1 to highly bqneficial impact.
5	Benthal flora	Impact -1 to +1	Discharge of industrial and fauna wastewater would affect banthal flora and fauna based on presence of nutrients, organic matter, heavy metals and toxic substances. In absence of prediction technique impact measurement would be subjective in which -1 corresponds to highly adverse and +1 to highly beneficial impact.
6	Algal growth	Impact -1 to +1	Discharge of industrial and fauna wastewater would affect algal growth and fauna based on presence of nutrients, organic matter, heavy metals and toxic substances. In absence of prediction technique impact measurement would be

			subjective in which -1 corresponds to highly adverse and +1 to highly beneficial impact
7	Fish yield	Impact -1 to +1	Discharge of industrial and fauna wastewater would affect algal growth and fauna based on presence of nutrients organic matter, heavy metals and toxic substances. In absence of prediction technique impact measurement would be subjective in which -1 corresponds to highly adverse and +1 to highly beneficial impact.

Category : Ecology
Component : Terrestrial ecology
Impact zone : 10 km. radial distance from the proposed unit

1	Green wer	Impact -1 to +1	Land acquisition for the project for industry as also dumping of wastes, and air pollutants emitted by the industry would affectthe green cover adversely. Activities such as compensatory afforestation, social forestry, horticulture etc. would have beneficial impact on green cover. In absence of prediction technique relating impact of pollution on vegetation, impact measurement would **be** subjective in which -1 corresponds to highly adverse and +1 to highly beneficial impact.
2	Crop yield	Impact -1 to +1	Air pollutants emitted by industry can have adverse impact due to oxides of sulphur while micronutrients present in flyash can have beneficial impact on crop yield. Similarly use of wastewater containning nitrogen and phosphorus can have beneficial impact on crop yield. In absence of prediction technique relating impact of pollution/ nutrients from industry on crop yield, impact measurement would be subjective in which -1 corresponds to highly adverse and +1 to highly beneficial impact.
3	Rare & endangered plant species	Impact -1 to +1	Air pollutants emitted by industry may affect rare and endangered species. However, it is not possible to quantify the impact, industry may initiate special efforts to preserve these species, impact of which again

			can not be quantified. Hence impact measurement would be subjective in which -I corresponds to highly adverse and +1 to highly beneficial impact
4	Rare & endangered animal species	Impact -1 to +1	Air pollutants emitted by industry may affect rare and endangered species. However, it is not possible Lo quantify the impact, industry may initiate special efforts to preserve these species, impact of which again can not be quantified. Hence impact measurement would be subjective in which -1 corresponds to highly adverse and +1 to highly beneficial impact.
5	Vegetation diversity	Impact -1 to +1	Change in landuse as also air pollutants emitted by industry may affect vegetation diversity. However, it is not possi- ble to quantify the impact. Industry may initiate special efforts to introduce vegetation diversity through wasteland reclamation, impa'ct of which again car1 not be quantified. Hence impact measurement would be subjctive in which -1 corresponds to highly adverse and +1 to highly beneficial impact.
6	Fauna diversity	Impact -1 to +1	Change in landuse as also air pollutants emitted by industry may affect vege- tation diversity. However, it is not possi- ble to quantify the impact. Industry may initiate special efforts to introduce fauna diversity through forest management, impact ofwhich again can not be quan- tified. Hence impact measurement would be subjctive in which -1 corres- ponds to highly adverse and +l to highly beneficial impact.
	Category : Environmental Pollution Component : air Impact Zone : 10 km radial distance from the proposed unit		
1	Suspended particulate	ug/cum	Maximum **24** - hrs concentration at the sensitive receptor
2	Oxides of sulphur	ug/cum	Maximum **24** - hrs concentration at the sensitive receptor
3	Oxides of nitrogen	ug/cum	Maximum **24** - hrs concentration at the sensitive receptor

4	Carbon monoxide	Impact -1 to +1	Carbon monoxide is mainly contributed by line sources which are not rnodelled in present research. Hence, impact me- asurement is subjective in which -1 corr- esponds to highly adverse and +1 to highly beneficial impact. Beneficial im- pact can be due to development of gre- en belt and traffic diversionlnew roads.
5	Total hydrocarbons	Impact -1 to +1	Hydrocarbons are mainly contributed by line sources and fugitive emissions which are not lnodelled in present rese- arch. Hence, impact measurement is subjective in which -1 corresponds to highly adverse and +1 to highly bene- ficial impact. Beneficial impact can be due to traffic diversionlnew roads.
6	Phytochemical oxidants		Photochemical oxidants are not model- led in the present research. Hence, impact measurement is subjective in which -1 corresponds to highly adverse and +1 to highly beneficial impact. Beneficial impact can be due to development of green belt.
	Category : Environmental Pollution Component : Water Impact Zone : Wastewater receiving waterbody		
1	dissoved oxygen	mg/lit	Minimum concentration downstream of the wastewater discharge point of the proposed project
2	Biochemical oxygen demand	mg/lit	Maximum concentration downstream of the wastewater discharge point of the proposed project
3	PH	deviation from 7.0	Maximum deviation downstream of from 7.0 the wastewater discharge point of the proposed project
4	Temperature	Increase in oC	Maximum increase in temperature downstream of the wastewater discharge point of the proposed project
5	Faecal coliform	MPNI100 ml	Maximum concentration downstream of the wastewater discharge point of the proposed project
6	Total Phosphates	mg/1	Maximum concentration downstream of the wastewater discharge point of the proposed project
7	Ammonia nitrogen	mg/1	Maximum concentration downstream of the wastewater discharge point of the proposed project-0

8	Turbidity	Impact -1 to +1	Impact on turbidity downstream of the wastewater discharge point of the proposed project. Parameter measure-ement is subjective in which -1 relates to significant increase in turbidity and +1 to significantdecrease in turbidity
9	Total dissolved solids	mg/l	Maximum concentration downstream of the wastewater discharge point of the proposed project
	Category : Environmental Pollution Compact : Land Impact Zone : Waste disposal area		
1	Exchangeable sodium %	Impact -1 to +1	If exchangeable sodium Percentage is more than 15.0 then soil becomes impermiable & sticky when wet and very hard when dry. ESP is calculated as ratio of concentration of sodium to total concentration of sodium, potassium, calcium and magnesium. In absence of prediction technique impact measurement is subjective in which-1 corresponds to highly adverse and +1 to highly beneficial impact. Beneficial impact can be due soil conditioning nature of sludgdwastewater.
2	Water extractable metals	Impact -1 to +1	If soil contains water extractable metals and if project aims at disposal of wastewater on land then if may lead to pollution of ground and surface water sources. In present research impact measurement is subjective in which-1 corresponds to highly adverse and +1 to highly beneficial impact. Beneficial impact can be due soil conditioning nature of sludg/wastewater.
3	Soil leachate quality	Impact -1 to +1	Discharge of industrial wastewater/ sludge will affect soil leachate quality. In present research impact measure-ment is subjective in which-1 corresponds to highly adverse and +1 to highly beneficial impact. Beneficial impact can be due soil conditioning nature of sludge/wastewater.
	Category : Environmental Pollution Component : Noise Impact zone : 2 km radial distance from the proposed unit		
1	Noise Level -	max.dB	Noise level in the immediate vicinity of the proposed site say within a distance of 2 km.

2	Noise Level - frequent	max.dB	Noise level in the i~nnlediatev icinity of the proposed site say within **a** distance of **2** km.
3	Noise Level - continuous	max.dB	Noise level in the immediate vicinity of the proposed site say within a distance of 2 km.

Category: Aesthetics
Component : Aesthetics
Impact zone : **2** km. radial distance from the pruposed unit

1	Landscape	Impact -1 to +l	Industrial project related constru-ctions may affect the landscape subsantially. Impact measurement is subjective in which -1 corresponds to highly adverse and +I to highly beneficial impact. Beneficial impact can be due to archtecture and greenbelt.
2	Appearance of waterbody	Impact -1 to +1	Parameter is relevant only when waste is discharge in waterbody of abstraction of water results in change in waterprofile. lmpact measurement is subjective in which -1 corresponds to highly adverse and +1 to highly beneficial impact. Beneficial impact can be due to archtecture and greenbelt.
3	Odour nuisance	Impact -1 to +1	Industrial production activities as also waste management facilities may cause adour nuissance for nearby inhabitatns. Impact measu-rement is subjective in which -1 corres-ponds to highly adverse and +1 to highly beneficial impart. Bieneficial impact can be due to archtecture and greenbelt
4	Fly nuisance	Impact -1 to +1	Industrial production activities as also waste management facilities may cause fly nuissance to nearby inhabitants. lmpact measurement is subjective in which -1 corresponds to highly adverse and +1 to highly beneficial impact. Beneficial impact can be due to archkcture & greenbelt.
5	Visibility	Impact -1 to +1	Emission of suspended particulate matter fro111 surface and elevated sources may lead to reduction in visibility specially suring winter season near the project site. Impact tneasurement is subjective in which -1 corresponds to highly adverse and +1 to highly beneficial impact

			Beneficial impact can be due to archtecture and greenbelt.
6	Composite Effect	Impact -1 to +1	Impact measurement is subjective in which -1 corresponds to highly adverse and +1 to highly beneficial impact. Beneficial impact can be due to archtecture and greenbelt.

Category: Human Interest
Component : Human Interest
Impact zone : 5 km. radial distance from the proposed unit.

1	Housing facilities	Impact -1 to +1	Workforce employed in the industry would require housing facilities near the project site. Unless township is developed as a part of the project this would result in additional demand on existing housing facilities. Impact measurement is subjective in which -1 corresponds to highly adverse and +1 to highly beneficial impact. Beneficial impact can be due to welfare housing schemes.
2	Employment opportunities		Industrial project would result in employment opportunities thus leading to improvement in socio-economic conditions. Parameter measurement is total number of jobs opportunities made available due all project related activities
3	Health facilities	Impact -1 to +1	As a welfare activity industry may develop health facilities doctor ratio resulting in beneficial impact. Otherwise there would be stress on existing facilities due to imigration. Impact measurement is subjective in which -1 corresponds to highly adverse and +1 to highly beneficial impact. Beneficial impact can be due to opening of industrial health facilities to general public.
4	Sanitation facilities	Impact -1 to +1	As a welfare activity industry may facilities develop health facilities doctor ratio resulting in beneficial impact. Otherwise there would be stress on existing facilities due to imigration. Impact measurement is subjective in which -1 corresponds to highly adverse and +1 to highly beneficial impact. Beneficial impact can be due to welfare sanitation measures.

5	Educational facilities	Impact -1 to +1	As a welfare activity Industry may develop. educational facilities resulting in beneficial impact. Otherwise student to teacher ratio would further increase due to imigration thus adversely affecting existing facilities. Impact measurement is subjective in which -1 corresponds to highly adverse and +1 to highly beneficial impact.
6	Population Density	Impact -1 to +1	Imigration of workers would result in increase in gross population square krn. density which in turn may lead to congestion. Impact measurement is subjective in which -1 corresponds to highly adverse and +1 to highly beneficial impact. Beneficial impact can be due to satellite township.
7	Public Transport	Impact -1 to +1	Industrial activity may affect existing public transport facilities due to new jobs. Impact measurement is subjective in which -1 corresponds to highly adverse and +1 to highly beneficial impact. Beneficial impact can be due to provision public transport by industry.
8	Drinking water supply	Impact -1 to +1	Water requirement of industry may affect drinking water supply unless separate provisions are made. Impact measurement is subjective in which -1 corresponds to highly adverse and +l to highly beneficial impact. Beneficial impact can be due to welfare housing schemes.
9	Power supply	Impact -1 to +1	Industrial power requirement forms one of the competitive uses of and hence allocation of the same to industry may affect other users. Impact measurement is subjective in which -1 corresponds to highly adverse and +1 to highly beneficial impact. Beneficial impact can be due to welfare electrification.
10	Perishable food	Impact -1 to +1	Immigration due to prices to industry would exert demand for perishable and nonrerishable food items which are supplied by surrounding rural areas. Impact measurement is subjective in which -1 corresponds to highly adverse and +1 to highly beneficial

			impact. Beneficial impact can be due to subsidized perishable food supply by industry.
13	Land Price	Impact -1 to +1	Development of industrial facilities may result in increase in cost of land. Impact measurement is subjective in which -1 corresponds to highly adverse and +1 to highly beneficial impact. Beneficial impact can be due to economic developmental prospects.

The Impact Evaluation System (IES) developed in the present research uses scaling-weighting checklist for impact evaluation. This warrants that environmental impacts are expressed in commensurate units providing common units for comparison. The steps involved in the development of commensurate units include transformation of parameter estimates into an environmental quality (EQ) scale, assignment of importance units to the individual parameters (PIU), and the multiplication of scale values by importance values to obtain environmental impact units (EIU) for baseline, and implementation of project without environmental management plan, and project with environmental management plan scenarios as:

$$EIU = \sum_{i} \sum_{j} \sum_{k} (EQ)ijk * (P1U)ijk \qquad (3.67)$$

where

i = category number (1 to 4)

j = component number in category i

k = parameter nuber in component (i,j)

Transformation of parqmeter estimates into an EQ scale is based on the range of anticipated value for a given parameter. To help transform the parameter estimates into an EQ scale, value function graphs have been developed showing parameter measurements on abscissa and EQ scale (0 to 1) on the ordinate. Assignment of importance units to each individual parameter is based on ranked pairwise comparison ,technique (Table **3.39**).

The overall computational methodology for impact evaluation involves following steps:

i. Select categories, components and parameters for impact evaluation from the comprehensive list inbuilt in the software. Only those parameters should be selected for which quantitative/qualitative predictions can be made.

Table 3.39. Numerical Database for value function curves for impact evaluation

S.N.	Parameter	Measurement Unit	Parameter Measurement corresponding to Environmental Quality										
			0.0	0.1	0.2	0.3	0.4	0.5	0.6	0.7	0.8	0.9	1.0
	Category : Ecology Component : Aquatic ecology												
1	River flow	impact -1 to 1	-1.0	-0.8	-0.6	-0.4	-0.2	0.0	0.2	0.4	0.6	0.8	1.0
2	Aq. nat. vegetation	impact -1 to 1	-1.0	-0.8	-0.6	-0.4	-0.2	0.0	0.2	0.4	0.6	0.8	1.0
3	Phytoplankton diversity	impact -1 to 1	-1.0	-0.8	-0.6	-0.4	-0.2	0.0	0.2	0.4	0.6	0.8	1.0
4	Zooplankton diversity	impact -1 to 1	-1.0	-0.8	-0.6	-0.4	-0.2	0.0	0.2	0.4	0.6	0.8	1.0
5	Benthal flora & fauna	impact -1 to 1	-1.0	-0.8	-0.6	-0.4	-0.2	0.0	0.2	0.4	0.6	0.8	1.0
6	Algal growth	impact -1 to 1	-1.0	-0.8	-0.6	-0.4	-0.2	0.0	0.2	0.4	0.6	0.8	1.0
7	Fish yield	impact -1 to 1	-1.0	-0.8	-0.6	-0.4	-0.2	0.0	0.2	0.4	0.6	0.8	1.0
	Category : Ecology Component : Terrestrial ecology												
1	Green cover	impact -1 to 1	-1.0	-0.8	-0.6	-0.4	-0.2	0.0	0.2	0.4	0.6	0.8	1.0
2	Crop yield	impact -1 to 1	-1.0	-0.8	-0.6	-0.4	-0.2	0.0	0.2	0.4	0.6	0.8	1.0
3	Rare & endangered species	impact -1 to 1	-1.0	-0.8	-0.6	-0.4	-0.2	0.0	0.2	0.4	0.6	0.8	1.0
4	Rare & endangered species	impact -1 to 1	-1.0	-0.8	-0.6	-0.4	-0.2	0.0	0.2	0.4	0.6	0.8	1.0
5	Vegetation diversity	impact -1 to 1	-1.0	-0.8	-0.6	-0.4	-0.2	0.0	0.2	0.4	0.6	0.8	1.0
6	Fauna diversity	impact -1 to 1	-1.0	-0.8	-0.6	-0.4	-0.2	0.0	0.2	0.4	0.6	0.8	1.0
	Category : Environmental Pollution Component : Air Environment												
1	susp. part. matter	ug/cum	500.0	300.0	225.0	150.0	125.0	100.0	75.0	50.0	25.0	10.0	0.0
2	Oxides of sulphur	ug/cum	1310.0	655.0	420.0	315.0	236.0	197.0	157.0	131.0	105.0	65.0	0.0

Table 3.39 contd...

3	Oxides of nitrogen	ug/cum	1230.0	209.0	80.0	61.0	55.0	50.0	37.0	34.0	30.0	18.0	0.0
4	Carbon monoxide	impact -1 to 1	-1.0	-0.8	-0.6	-0.4	-0.2	0.0	0.2	0.4	0.6	0.8	1.0
5	Total hydrocarbons	impact -1 to 1	-1.0	-0.8	-0.6	-0.4	-0.2	0.0	0.2	0.4	0.6	0.8	1.0
6	Photochemical oxidants	impact -1 to 1	-1.0	-0.8	-0.6	-0.4	-0.2	0.0	0.2	0.4	0.6	0.8	1.0
	Category : Environmental Pollution Component : Water Environment												
1	Dissolved oxygen	mg/l	0.0	2.0	3.0	3.5	4.2	5.0	5.5	6.0	6.5	7.5	9.0
2	Biochem. oxygen demand	mg/l	20.0	15.0	10.0	8.0	7.0	6.0	5.0	4.0	3.0	2.0	0.0
3	pH	impact -1 to 1	5.0	3.0	2.5	2.3	1.8	1.4	1.2	1.1	1.0	0.5	0.0
4	Temperature	increase 0C	10.0	5.0	4.5	4.0	3.5	3.0	2.5	1.8	1.0	0.5	0.0
5	Faecal coliform	mpn/100 ml	100000	10000	1000.0	200.0	100.0	30.0	20.0	10.0	5.0	3.0	0.0
6	Total phosphates	mg/l	10.0	5.0	2.5	1.5	1.0	0.8	0.6	0.4	0.2	0.1	0.0
7	Ammonia nitrogen	mg/l	100.0	55.0	35.0	22.0	15.0	10.0	7.5	5.0	2.5	1.0	0.0
8	Turbidity	impact -1 to 1	-1.0	-0.8	-0.6	-0.6	-0.2	0.0	0.2	0.4	0.6	0.8	0.0
9	Total dissolved solids	mg/l	2000.0	1500.0	1000.0	500.0	450.0	375.0	300.0	225.0	150.0	150.0	150.0
	Category: Environmental Pollution Compnent : Land Environment												
1	Exchangeable sodium %	impact -1 to 1	-1.0	-0.8	-0.6	-0.4	-0.2	0.0	0.2	0.4	0.6	0.8	1.0
2	Water extractable metals	impact -1 to 1	-1.0	-0.8	-0.6	-0.4	-0.2	0.0	0.2	0.4	0.6	0.8	1.0
3	Soil permiability	impact -1 to 1	-1.0	-0.8	-0.6	-0.4	-0.2	0.0	0.2	0.4	0.6	0.8	1.0
	Category : Environment Pollution Component : Noise Environment												
1	Max. noise level - sprace	db	100.0	95.0	90.0	85.0	80.0	75.0	70.0	65.0	60.0	55.0	50.0
2	Max. noise level- frequent	db	100.0	92.0	85.0	75.0	70.0	65.0	62.0	59.0	57.0	55.0	50.0

Table 3.39 contd...

3	Max. noise level-continuous	db	100.0	85.0	75.0	70.0	65.0	61.0	57.0	55.0	53.0	51.0	50.0
	Category : Aesthetics Component : Aesthetics												
1	Landscape	impact -1 to 1	-1.0	-0.8	-0.6	-0.4	-0.2	0.0	0.2	0.4	0.6	0.8	1.0
2	Appearance of water body	impact -1 to 1	-1.0	-0.8	-0.6	-0.4	-0.2	0.0	0.2	0.4	0.6	0.8	1.0
3	Odour nuissance	impact -1 to 1	-1.0	-0.8	-0.6	-0.4	-0.2	0.0	0.2	0.4	0.6	0.8	1.0
4	Odour nuissance	impact -1 to 1	-1.0	-0.8	-0.6	-0.4	-0.2	0.0	0.2	0.4	0.6	0.8	1.0
5	Visibility	impact -1 to 1	-1.0	-0.8	-0.6	-0.4	-0.2	0.0	0.2	0.4	0.6	0.8	1.0
6	Composite effect	impact -1 to 1	-1.0	-0.8	-0.6	-0.4	-0.2	0.0	0.2	0.4	0.6	0.8	1.0
	Category : Human Interest Component : Human Interest												
1	Housing facilities	impact -1 to 1	-1.0	-0.8	-0.6	-0.4	-0.2	0.0	0.2	0.4	0.6	0.8	1.0
2	Employment opportunities	nos.	0.0	100.0	200.0	300.0	400.0	500.0	600.0	700.0	800.0	900.0	1000.0
3	Health faciliteis	impact -1 to 1	-1.0	-0.8	-0.6	-0.4	-0.2	0.0	0.2	0.4	0.6	0.8	1.0
4	Sanitation facilities	impact -1 to 1	-1.0	-0.8	-0.6	-0.4	-0.2	0.0	0.2	0.4	0.6	0.8	1.0
5	Education facilities	impact -1 to 1	-1.0	-0.8	-0.6	-0.4	-0.2	0.0	0.2	0.4	0.6	0.8	1.0
6	Population density	impact -1 to 1	-1.0	-0.8	-0.6	-0.4	-0.2	0.0	0.2	0.4	0.6	0.8	1.0
7	Public transport facilities	impact -1 to 1	-1.0	-0.8	-0.6	-0.4	-0.2	0.0	0.2	0.4	0.6	0.8	1.0
8	Drinking water supply	impact -1 to 1	-1.0	-0.8	-0.6	-0.4	-0.2	0.0	0.2	0.4	0.6	0.8	1.0
9	Power supply	impact -1 to 1	-1.0	-0.8	-0.6	-0.4	-0.2	0.0	0.2	0.4	0.6	0.8	1.0
10	Perishable food prices	impact -1 to 1	-1.0	-0.8	-0.6	-0.4	-0.2	0.0	0.2	0.4	0.6	0.8	1.0
11	Land price	impact -1 to 1	-1.0	-0.8	-0.6	-0.4	-0.2	0.0	0.2	0.4	0.6	0.8	1.0

ii. Read data for all categories, components and parameters for baseline scenario, with project and without environmental management plan scenario,. and with project and environmental management plan scenario

iii. Assign total score of 1000 to categories, components and parameters through ranked pairwise comparision technique inbuilt in the programme

iv. For categories i = 1 to 4, repeat stepts (v) through (xi·)

v. For components j = 1 to (ij) repeat steps (vi) through (xi)

vi. Calculate Environmental Quality EQijk corresponding to parameter measurement (ijk) using value function curve

vii. Calculate Environmental Impact Units EIUijk for the parameter as summation of EQijk" PIUvijk

viii. Calculate EIU for the component as CEIUijk ix. Calculate EIU for the category as C EIUjk x. Calculate EIU for the alternative as C EIUk

ix. Calculate environmental indices (E1)j for individual component by deviding category EIU by component score

x. Calculate environmental indices (EI)k for individual category by deviding category EIU by category score

xi. Tabulate results (EIU) for individual categories and aggregate scores for the scenario

Calculate change in EIU between scenarios a & b, a & c, and b &. c

Impact evaluation module in EIA-AID incorporates the above. algorithm. Systems flowchart for impact evaluation is shown in Fig. 1.10.

Salient features of the IES module are :

- Assigning weightage to categories and components through ranked pairwise comparison technique in interactive mode
- Flexibility of modifying weights so assigned
- On screen selection of parameters and assigning weights
- Default weights assigned for naive user
- Value Function Curves stored for all parameters
- User can modify VFC as per his perception

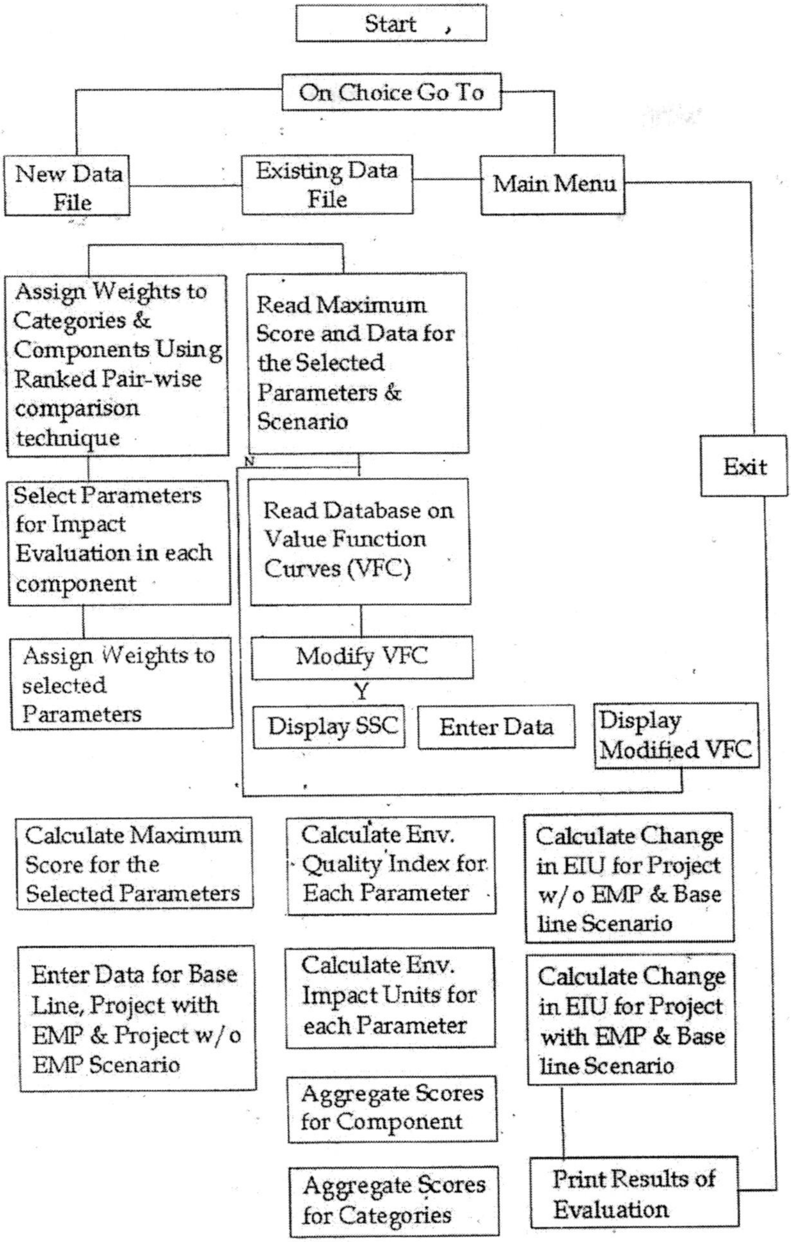

Fig. 3.10. System Flow-chart for IES : Impact Evaluation

- Option to modify scaling function database using text-editor
- Measurement units defined on screen for easy data entry
- lmpacd zones defined on screen for easy data entry
- On s'creen confirmation for correct data entry in interactive mode

3.7 SENSITIVITY ANALYSIS IN IMPACT EVALUATION

The resultant numerical evaluation in IES must be subjected to professional interpretation to provide analysis of trade-off within a component, within a category and between categories. The uncertainty in assigning scores to categories, components and parameters is the major limitation of IES which could be overcome through sensitivity analysis with respect to importance units (Dutt, 1993).

The sensitivity analysis methodology developed in this research involves following steps :

i. Read Environmental Indices (EI) for all categories for
 a. baseline scenario
 b. with project and without environmental management plan scenario
 c. with project and environmental management plan scenario

ii. Decide the threshold change in EIU (compared to baseline scenario) for project proposal to be termed as acceptable for environmental clearance

iii. Decide the step (numerical value for perceptible change) for-varying scores (say **50)**

iv. Generate combinations of scores that can be assigned to the individual categories (CIU)I

v. Calculate Environmental Impact Units EIUi for the categories as EIi'TIUi

vi. Aggregate EIUs for the scenario

vii. Calculate change in EIU between scenarios a & c as Net EIU

viii. Compute maximum score that can be assigned to a category so that Net EIU is more than threshold

Sensitivity analysis module in EIA-AID calculates project acceptancelrejection criteria for various environmental categories generating all possible combinations of PIUs for user defined step size incorporating the above algorithm.

3.8 DESIGN OF GREEN BELT

T'hc fugitive emissions and waste residuals subsequent to pollution control measures could be attenuated by the development of a green belt around the industry. Green belts not only absorb air and water pollutants but also help in arresting noise, and soil erosion, and creating favourable aesthetic conditions.

Design of green belt involves estimation of the width of green belt based on the meteorolng-ical conditions, pollutants to be attenuated, desired level of pollution attenuation, and the nature (height, canopy cover, leaf area index) of the plant species available locally.

The effectiveness of a green belt in attenuating the pollution is given by the attenuation factor **AF,** which is defined as the ratio of mass flux of pollutant reaching at a distance in absence of green belt to the mass flux reaching at the same distance in presence of green belt (Kapoor and Gupta, 1984).

Computation of pollution attenuation due to green-belt to decide optimal width in this research involves following steps:

i. Read data for pollutant to be attenuated, dry deposition velocity of pollutant for the plant species selected average foliage surface area density for the plant species, individual folliage surface area density, Pasquil- Gifford atmospheric stability condition, wind speed outside green belt, wind speed inside green belt, height pollutant release and distance of green belt begining from the source

ii. Estimate pollution attenuation coefficient for the plant species

iii. Estimate mass flux of pollutants reaching the inner edge of green belt

iv. Repeat steps (v) through (x) for different combinations of green belt width and height

v. Estimate fraction of mass flux entering the green belt for different average heights of green belt

vi. Estimate mass flux of pollutants coming out of green belt

vii. calculate attenuation factor

viii. Calculate pollutant concentration at outer edge of green belt without green belt using Gaussian Plume Model

ix. Calculate pollutant concentration at outer edge of green belt with green belt accounting for attenuation

x. Go to step (v) if change in attenuation is more than 1 'percent, else stop

Incorporating this algorithm and the mathematical formulation suggested by Kapoor and Gupta, Green belt design module in EIA-AID estimates attenuation factors and concentration at the receptor located outside the green-belt with and without green-belt to decide the optimal width of green belt based on sensitivity of the receptor.

REFERENCES

APHA (1987). Standard Methods of the Examination of Water and Wastewater (17th ed.) American Public Health Association, Washington D.C.

Barnwell T.M., Jr., Brown, L.C., and Whitemore, R.C. (1987). QUAL- 2E : **A** Case Study in Water ~ u a l i t !M~o deling Software in Beck M.B. ed. Systems Analysis in Water Quality Management, IAWPRC, Pergamon Press, New York

Bathala, C.T., Das, K.C. and Jones, W.D. (1979). Assimilative Capacity of Small Streams, Journal of Environmental Engineering Division, ASCE, Vol. 105, No. EE6, pp. 1049-1062.

Briggs, G.A. (1973). Diffusion estimation for small emissions, AIDD contribution 79, Air Resources, Atmospheric Turbulent and Diffusion Laboratory, Oak Ridge, T.N.

Briggs, G.A. (1975). Plume rise predictions, lectures on Air Pollution and Environmental Impact Analysis, American Met. Society, Boston.

Camp, T.R. (1963). Water and its Impurities, Chapman and Hall, London.

Carpenter, S. B., Montgomey, T. L., Leauitt, J. M., Colbaugh W. C. and Thomas, F. W. (1971). Principal Plume Dispersion Models, TVA Power Plants, Journal of Air Pollution Control Association, Vol. 21, pp. 491-495

Churchill, M.A., Elmore, H.L., and Buckingham, A.A. (1962). The Prediction of Stream Reaeration Rates, Journal of the Sanitary Engineering Division, ASCE, Vol. 88, No. SA4, July 1962, pp. 1- 46.

Constable, T.W. and McBean, E.A. (1979). Kalman Filter Modelling of th Speed River Quality Journal of Environmental Engineering Division, ASCE, vol. 105, No. EE5, pp. 961-978.

Dhokalia, R.P. (1993). The Quality of Life, paper presented in Workshop on National Programme for Environmental Management for Coal Fired Power Generation, May 12-13, Ministry of Environment and Forests, New Delhi

EPA (1976). Compilation of Air Pollution Emission Factors : AP-42, US Environmental Protection Agency

Dobbins, W.E. (1964). BOD and Oxygen Relationships in Streams, Journal of Sanitary Engineering Division, ASCE, Vol. 90, No. SA3, pp. 53-78.

Flavin, C. and Lenssen, N. (1994). Reshaping the Power Industry in State of the World : 1994, A Worldwatch Institute Report, W.W. Norton & Company, New York

Gifford, F. A. (1961). Use of Routine Meteriological Observations for Estimating Atmospheric Dispersion, Nuclear Safety, Vol. 2 No. 4 pp 47-57

Gifford, F.A. and Hanna, S.R. (1973). Modelling Urban Air Pollution,

Atmos. Environment, 7, 131-136.

Gromiec, M., Louks, D.P., and Orlob, G.T. (1983). Stream Quality Modelling in Mathematical Modeling of Water Quality: Streams, Lakes, and Reservoirs edited by Orlob, G.T, John Wiley & Sons, New-York, pp. 176-226.

Irwin, J.S. (1979). A theoretical variation of the wind profile power law experiment as a function of source roughness and stability, Atm. Environment, 13, 191-194.

IS1 (1978). Guidelines for Micrometeorological Techniques in Air Pollution Studies, IS:8829-1978, Indian Standards Institution, New Delhi 110 002, India.

Kapoor, S. (1992). Sources of Noise Pollution in Oil Exploring and Exploiting Industry - Prevention and Control in Wahi S. K ***et a1*** eds. Environmental Management in Petroleum Industry, Wiely Eastern Limited

Kapoor, R.K. and Gupta, V.K. (1984). A Pollution Attenuation Coefficient Concept for Optimization of Green Belt, Journal of Atmospheric Environment, Vol. 18, No. 6? pp. 1107-113.

Lucas, D.H. (1967). Application and Evaluation of Results of the Tilbury Plume Rise and Dispersion Experiment, Atmos. Environ., 1, 421-424.

Luna, R.E. and Church, H.W. (1972). A Comparison of Turbulence Intensity and Stability Ratio Measurement to Pasquill Classes, J. Appl. Meteor, 11(4), 663-669.

Madhusudan Rao, P. (1991). Noise Pollution - Control Technology and Instruments in Industrial Safety and Pollution Control Handbook published by National Safety Council, India

MEF , (1985). Environment:) 1 Guidelines for Siting of Industrial Projects, Ministry of E1i\.i';onrnent and Forests, Government of India

MEF (1989). Gazette Notification dated 26 December 1989 under the Environment Protection Act 1986 Published by Ministry of Environment and Forests, Government of India

MEF (1993a). Gazette Notification GSR 422(E) dated 19 May 1993 under the Environment Protection Act 1986 Published by Ministry of Environment and Forests, Government of India

MEF (1993b). Gazette Notification No. 80(E), 28 January 1993, Ministry of Environment and Forests, Government of India

Mishra, U. C. and Mahadevan, T.N. (1991). Measurement Norms and Techniques for Environmental Quality - Air and Noise in Industrial Safety and Pollution Control Handbook published by National Safety Council, India

Miller, J. E., Jennings, M. E. '(1979). Modeling Nitrogen, Oxygen, Chattahoochee River, Ga,, Journal of Environmental Engineering Division, ASCE, Vol. 105, No. EE4, pp. 641-654

O'Connor, D.J., and Dobbins, W.E. (1958). Mechanism of Reaeration in Natural Streams Transactions of ASCE, paper no. 2934, pp. 641- 684

Owens, M., Edwards, R.W., an d Gibbs, J.W. (1964). Some Reaeration Studies in Streams, International Journal of Air and Water Polluton, vol. 8, pp. 469-486

Pawlow, J.R. (1983). Technical Guidance Manual for Performing Waste load Alloacations, Book 11, Streams and Rivers, US-EPA, Sept. 1983

Pasquill, F. (1961). The estimation of the dispersion of windborne material, Meteoral. Mag., 90, 33-49

Pasquill, F. (1974). Atmospheric Diffusiob, 2nd ed., John Wiley & Sons, New York, 429 .

Rau, J.G. and Wooten, D.C. (1980). Environmental Impact Analysis Handbook, McGraw-Hill Book Company, New York

Smith (Ed) M.E. (1968). Recommended Guide for the prediction of the dispersion of airborne effluents, 1st ed., ASME, New York

Sangar, C.B.S. (1993). Noise Pollution : Measurement and Control, Indian Journal of Environmental Protection, Vol. 13, No. 7, pp. 517 -520

Streeter, H.W. and Phelps, E.B. (1925). A Study of Pollution and Natural Purification of the Ohio River, U.S. Public Health Services, Washington D.C., Bulletin 146.

Sharma, S.C. (1993). Prediction Techniques in Environmental Impact Assessment : Air Quality, paper presented in Workshop on Methodologies and Procedures in Environmental Impact

Assessment of Industrial and River Valley Projects, November 22-26, NEERI, Nagpur, India

Strauss, W. (1980). Formation and Control of Air Pollutants in Bockris J. O.M., ed., Environmental Chemistry, Plenum Press, New York

Thomann, R.V. (1987). Systems Analysis in Water Quality Management a 25 year Retrospect, in Beck M.B. ed. Systems Analysis in Water Quality Management, IAWPRC, Pergamon Press, New York

Thomas, H.A. (1948). Pollution Load Capacity of Streams, Water and Sewage Works, Vol. 95, pp. 409-417.

Turner, D.B. (1970). Work book of Atmospheric Dispersion Estimates, US public Health Service, publication 999-AP-26, Cinicinmati, Ohio

Wright, R.M., and McDonell, A.J. (1979). In-stream Deoxygenation Rate Prediction, Journal of Environmental Engineering Division, ASCE, Vol. 105, No. EE2, pp. 323-335.

CHAPTER - 4

DISCUSSIONS AND CONCLUSIONS

The present research was addressed to development of methodology for EIA of industrial projects, computerization of the methodology and validation of the same through a real life case study. The salient contributions of the present research to development of methodology for EIA of industrial projects relate to:

- Development of methodology for screening of industrial projects based on project thresholds
- Development of methodology based on premises of carrying capacity based developmental planning in ranking of site alternatives of industrial projects, and scaling weighing checklist incorporating development of site acceptability functions for 73 parameters related to ecology and economics
- Development of knowledgebase for impact identification, baseline data collection, and prediction techniques
- Development of methodology for impact identification using questionnaire checklist
- Development of prediction models for air, water and noise components incorporating state-of-the art mathematical models and expert guesstimates for biological and socio-economic
- Sensitivity analysis of the prediction models to assist scoping for baseline data collection
- Development of impact evaluation system combining the best features of Leopold matrix and scaling weighing checklist

- Development of methodology for sensitivity analysis in impact evaluation

The salient features of software package developed in the present research, EIA-AID, relate to :

- EIA-AID is a stand alone package which can run under DOS environment on PC platform
- **As** the total size of package is **2** MB it can be zipped and transported/disseminated
- EIA-AIr has been developed in QUICKBASIC 4.5. Package provides pleasant screel~d isplays
- While the graphics por ,ion in ROSA and IES module requires EGA monitor, remaining package can be used even with monochrome monitor
- Data entry is user friendly and on-line help has been provided
- Out put of EIA-AID is self explanatory and can be used directly in EIA report

The methodology and the software package developed in the present research is a unique package designed specifically for the EIA of industrial projects is expected to simplify the process of EIA report preparation and appraisal. Use of EIA-AID in the process of EIA can make the decision making transparent and eliminate scope of engineered EIA. Also, in view of interdisciplinary nature of EIA study and appraisal, EIA-AID would be extremely useful tool for enlightened participation in EIA.